THE PRINCIPLE OF RELATIVITY

A COLLECTION OF ORIGINAL MEMOIRS ON THE SPECIAL AND GENERAL THEORY OF RELATIVITY

BY

H. A. LORENTZ, A. EINSTEIN
H. MINKOWSKI AND H. WEYL

WITH NOTES BY

A. SOMMERFELD

TRANSLATED BY

W. PERRETT AND G. B. JEFFERY

WITH SEVEN DIAGRAMS

DOVER PUBLICATIONS, INC.

Published in Canada by General Publishing Company, Ltd., 30 Lesmill Road, Don Mills, Toronto, Ontario.

Published in the United Kingdom by Constable and Company, Ltd., 10 Orange Street, London WC 2.

This Dover edition, first published in 1952, is an unabridged and unaltered republication of the 1923 translation first published by Methuen and Company, Ltd. in 1923. This edition is reprinted through special arrangement with Methuen and Company and Albert Einstein.

Standard Book Number: 486-60081-5
Library of Congress Catalog Card Number: A52-9845

Manufactured in the United States of America
Dover Publications, Inc.
180 Varick Street
New York, N.Y. 10014

TRANSLATORS' PREFACE

THE Theory of Relativity is at the moment the subject of two main lines of inquiry: there is an endeavour to express its principles in logical and concise form, and there is the struggle with analytical difficulties which stand in the way of further progress. In the midst of such problems it is easy to forget the way in which the theory gradually grew under the stimulus of physical experiment, and thus to miss much of its meaning. It is this growth which the present collection of papers is designed chiefly to exhibit. In the earlier papers there are some things which the authors would no doubt now express differently; the later papers deal with problems which are not by any means yet fully solved. At the end we must confess that Relativity is still very much of a problem—and therefore worthy of our study.

The authors of the papers are still actively at work on the subject—all save Minkowski. His paper on "Space and Time" is a measure of the loss which mathematical physics suffered by his untimely death.

The translations have been made from the text, as published in a German collection, under the title "Des Relativitatsprinzip" (Teubner, 4th ed., 1922).

The second paper by Lorentz is an exception to this. It is reprinted from the original English version in the Proceedings of the Amsterdam Academy. Some minor changes have been made, and the notation has been brought more nearly into conformity with that employed in the other papers.

W. P.
G. B. J.

TABLE OF CONTENTS

MICHELSON'S INTERFERENCE EXPERIMENT

BY

H. A. LORENTZ

Translated from "Versuch einer Theorie der elektrischen und optischen Erscheinungen in bewegten Körpern," Leiden, 1895, §§ 89-92.

MICHELSON'S INTERFERENCE EXPERIMENT

By H. A. LORENTZ

1. AS Maxwell first remarked and as follows from a very simple calculation, the time required by a ray of light to travel from a point A to a point B and back to A must vary when the two points together undergo a displacement without carrying the ether with them. The difference is, certainly, a magnitude of second order; but it is sufficiently great to be detected by a sensitive interference method.

The experiment was carried out by Michelson in 1881.* His apparatus, a kind of interferometer, had two horizontal arms, P and Q, of equal length and at right angles one to the other. Of the two mutually interfering rays of light the one passed along the arm P and back, the other along the arm Q and back. The whole instrument, including the source of light and the arrangement for taking observations, could be revolved about a vertical axis; and those two positions come especially under consideration in which the arm P or the arm Q lay as nearly as possible in the direction of the Earth's motion. On the basis of Fresnel's theory it was anticipated that when the apparatus was revolved from one of these *principal positions* into the other there would be a displacement of the interference fringes.

But of such a displacement—for the sake of brevity we will call it the Maxwell displacement—conditioned by the change in the times of propagation, no trace was discovered, and accordingly Michelson thought himself justified in concluding that while the Earth is moving, the ether does not remain at rest. The correctness of this inference was soon brought into question, for by an oversight Michelson had

* Michelson, American Journal of Science, 22, 1881, p. 120.

taken the change in the phase difference, which was to be expected in accordance with the theory, at twice its proper value. If we make the necessary correction, we arrive at displacements no greater than might be masked by errors of observation.

Subsequently Michelson * took up the investigation anew in collaboration with Morley, enhancing the delicacy of the experiment by causing each pencil to be reflected to and fro between a number of mirrors, thereby obtaining the same advantage as if the arms of the earlier apparatus had been considerably lengthened. The mirrors were mounted on a massive stone disc, floating on mercury, and therefore easily revolved. Each pencil now had to travel a total distance of 22 meters, and on Fresnel's theory the displacement to be expected in passing from the one principal position to the other would be 0·4 of the distance between the interference fringes. Nevertheless the rotation produced displacements not exceeding 0 02 of this distance, and these might well be ascribed to errors of observation.

Now, does this result entitle us to assume that the ether takes part in the motion of the Earth, and therefore that the theory of aberration given by Stokes is the correct one? The difficulties which this theory encounters in explaining aberration seem too great for me to share this opinion, and I would rather try to remove the contradiction between Fresnel's theory and Michelson's result. An hypothesis which I brought forward some time ago,† and which, as I subsequently learned, has also occurred to Fitzgerald,‡ enables us to do this. The next paragraph will set out this hypothesis.

2. To simplify matters we will assume that we are working with apparatus as employed in the first experiments, and that in the one principal position the arm P lies exactly in

* Michelson and Morley, American Journal of Science, 34, 1887, p. 333; Phil. Mag., 24, 1887, p. 449.

† Lorentz, Zittingsverslagen der Akad. v. Wet. te Amsterdam, 1892-93, p. 74.

‡ As Fitzgerald kindly tells me, he has for a long time dealt with his hypothesis in his lectures. The only published reference which I can find to the hypothesis is by Lodge, "Aberration Problems," Phil. Trans. R.S., 184 A, 1893.

the direction of the motion of the Earth. Let v be the velocity of this motion, L the length of either arm, and hence 2L the path traversed by the rays of light. According to the theory,* the turning of the apparatus through 90° causes the time in which the one pencil travels along P and back to be longer than the time which the other pencil takes to complete its journey by

$$\frac{Lv^2}{c^3}.$$

There would be this same difference if the translation had no influence and the arm P were longer than the arm Q by $\frac{1}{2}Lv^2/c^2$. Similarly with the second principal position.

Thus we see that the phase differences expected by the theory might also arise if, when the apparatus is revolved, first the one arm and then the other arm were the longer. It follows that the phase differences can be compensated by contrary changes of the dimensions.

If we assume the arm which lies in the direction of the Earth's motion to be shorter than the other by $\frac{1}{2}Lv^2/c^2$, and, at the same time, that the translation has the influence which Fresnel's theory allows it, then the result of the Michelson experiment is explained completely.

Thus one would have to imagine that the motion of a solid body (such as a brass rod or the stone disc employed in the later experiments) through the resting ether exerts upon the dimensions of that body an influence which varies according to the orientation of the body with respect to the direction of motion. If, for example, the dimensions parallel to this direction were changed in the proportion of 1 to $1 + \delta$, and those perpendicular in the proportion of 1 to $1 + \epsilon$, then we should have the equation

$$\epsilon - \delta = \frac{1}{2}\frac{v^2}{c^2} \qquad \qquad (1)$$

in which the value of one of the quantities δ and ϵ would remain undetermined. It might be that $\epsilon = 0$, $\delta = -\frac{1}{2}v^2/c^2$, but also $\epsilon = \frac{1}{2}v^2/c^2$, $\delta = 0$, or $\epsilon = \frac{1}{4}v^2/c^2$, and $\delta = -\frac{1}{4}v^2/c^2$.

3. Surprising as this hypothesis may appear at first sight,

* Cf. Lorentz, Arch. Néerl., 2, 1887, pp. 168-176.

yet we shall have to admit that it is by no means far-fetched, as soon as we assume that molecular forces are also transmitted through the ether, like the electric and magnetic forces of which we are able at the present time to make this assertion definitely. If they are so transmitted, the translation will very probably affect the action between two molecules or atoms in a manner resembling the attraction or repulsion between charged particles. Now, since the form and dimensions of a solid body are ultimately conditioned by the intensity of molecular actions, there cannot fail to be a change of dimensions as well.

From the theoretical side, therefore, there would be no objection to the hypothesis. As regards its experimental proof, we must first of all note that the lengthenings and shortenings in question are extraordinarily small. We have $v^2/c^2 = 10^{-8}$, and thus, if $\epsilon = 0$, the shortening of the one diameter of the Earth would amount to about 6·5 cm. The length of a meter rod would change, when moved from one principal position into the other, by about $\frac{1}{200}$ micron. One could hardly hope for success in trying to perceive such small quantities except by means of an interference method. We should have to operate with two perpendicular rods, and with two mutually interfering pencils of light, allowing the one to travel to and fro along the first rod, and the other along the second rod. But in this way we should come back once more to the Michelson experiment, and revolving the apparatus we should perceive no displacement of the fringes. Reversing a previous remark, we might now say that the displacement produced by the alterations of length is compensated by the Maxwell displacement.

4. It is worth noticing that we are led to just the same changes of dimensions as have been presumed above if we, *firstly*, without taking molecular movement into consideration, assume that in a solid body left to itself the forces, attractions or repulsions, acting upon any molecule maintain one another in equilibrium, and, *secondly*—though to be sure, there is no reason for doing so—if we apply to these molecular forces the law which in another place * we deduced for

* Viz., § 23 of the book, "Versuch einer Theorie der elektrischen und optischen Erscheinungen in bewegten Körpern."

electrostatic actions. For if we now understand by S_1 and S_2 not, as formerly, two systems of charged particles, but two systems of molecules—the second at rest and the first moving with a velocity v in the direction of the axis of x—between the dimensions of which the relationship subsists as previously stated; and if we assume that in both systems the x components of the forces are the same, while the y and z components differ from one another by the factor $\sqrt{1 - v^2/c^2}$, then it is clear that the forces in S_1 will be in equilibrium whenever they are so in S_2. If therefore S_2 is the state of equilibrium of a solid body at rest, then the molecules in S_1 have precisely those positions in which they can persist under the influence of translation. The displacement would naturally bring about this disposition of the molecules of its own accord, and thus effect a shortening in the direction of motion in the proportion of 1 to $\sqrt{1 - v^2/c^2}$, in accordance with the formulæ given in the above-mentioned paragraph. This leads to the values

$$\delta = -\tfrac{1}{2}\frac{v^2}{c^2}, \qquad \epsilon = 0$$

in agreement with (1).

In reality the molecules of a body are not at rest, but in every "state of equilibrium" there is a stationary movement. What influence this circumstance may have in the phenomenon which we have been considering is a question which we do not here touch upon; in any case the experiments of Michelson and Morley, in consequence of unavoidable errors of observation, afford considerable latitude for the values of δ and ϵ.

ELECTROMAGNETIC PHENOMENA IN A SYSTEM MOVING WITH ANY VELOCITY LESS THAN THAT OF LIGHT

BY

H. A. LORENTZ

Reprinted from the English version in Proceedings of the Academy of Sciences of Amsterdam, 6, 1904.

ELECTROMAGNETIC PHENOMENA IN A SYSTEM MOVING WITH ANY VELOCITY LESS THAN THAT OF LIGHT

By H. A. LORENTZ

§ 1. THE problem of determining the influence exerted on electric and optical phenomena by a translation, such as all systems have in virtue of the Earth's annual motion, admits of a comparatively simple solution, so long as only those terms need be taken into account, which are proportional to the first power of the ratio between the velocity of translation v and the velocity of light c. Cases in which quantities of the second order, i.e. of the order v^2/c^2, may be perceptible, present more difficulties. The first example of this kind is Michelson's well-known interference-experiment, the negative result of which has led Fitzgerald and myself to the conclusion that the dimensions of solid bodies are slightly altered by their motion through the ether.

Some new experiments, in which a second order effect was sought for, have recently been published. Rayleigh * and Brace † have examined the question whether the Earth's motion may cause a body to become doubly refracting. At first sight this might be expected, if the just mentioned change of dimensions is admitted. Both physicists, however, have obtained a negative result.

In the second place Trouton and Noble ‡ have endeavoured to detect a turning couple acting on a charged condenser, the plates of which make a certain angle with the direction of translation. The theory of electrons, unless it be modified by some new hypothesis, would undoubtedly require the

* Rayleigh, Phil. Mag. (6), 4, 1902, p. 678.

† Brace, Phil. Mag. (6), 7, 1904, p. 317.

‡ Trouton and Noble, Phil. Trans. Roy. Soc. Lond., A 202, 1903, p. 165.

existence of such a couple. In order to see this, it will suffice to consider a condenser with ether as dielectric. It may be shown that in every electrostatic system, moving with a velocity **v**,* there is a certain amount of "electromagnetic momentum." If we represent this, in direction and magnitude, by a vector **G**, the couple in question will be determined by the vector product †

$$[\mathbf{G} \, . \, \mathbf{v}] \quad . \quad . \quad . \quad . \quad . \quad (1)$$

Now, if the axis of z is chosen perpendicular to the condenser plates, the velocity **v** having any direction we like; and if U is the energy of the condenser, calculated in the ordinary way, the components of **G** are given ‡ by the following formulæ, which are exact up to the first order,

$$G_x = \frac{2U}{c^2} v_x, \quad G_y = \frac{2U}{c^2} v_y, \quad G_z = 0.$$

Substituting these values in (1), we get for the components of the couple, up to terms of the second order,

$$\frac{2U}{c^2} v_y v_z, \quad - \frac{2U}{c^2} v_x v_z, \quad 0.$$

These expressions show that the axis of the couple lies in the plane of the plates, perpendicular to the translation. If a is the angle between the velocity and the normal to the plates, the moment of the couple will be $U(v/c)^2 \sin 2a$; it tends to turn the condenser into such a position that the plates are parallel to the Earth's motion.

In the apparatus of Trouton and Noble the condenser was fixed to the beam of a torsion-balance, sufficiently delicate to be deflected by a couple of the above order of magnitude. No effect could however be observed.

§ 2. The experiments of which I have spoken are not the only reason for which a new examination of the problems connected with the motion of the Earth is desirable. Poin-

* A vector will be denoted by a Clarendon letter, its magnitude by the corresponding Latin letter.

† See my article: "Weiterbildung der Maxwell'schen Theorie. Electronentheorie," Mathem. Encyclopädie, V, 14, § 21, a. (This article will be quoted as "M.E.")

‡ "M.E.," § 56, c.

caré * has objected to the existing theory of electric and optical phenomena in moving bodies that, in order to explain Michelson's negative result, the introduction of a new hypothesis has been required, and that the same necessity may occur each time new facts will be brought to light. Surely this course of inventing special hypotheses for each new experimental result is somewhat artificial. It would be more satisfactory if it were possible to show by means of certain fundamental assumptions and without neglecting terms of one order of magnitude or another, that many electromagnetic actions are entirely independent of the motion of the system. Some years ago, I already sought to frame a theory of this kind.† I believe it is now possible to treat the subject with a better result. The only restriction as regards the velocity will be that it be less than that of light.

§ 3. I shall start from the fundamental equations of the theory of electrons.‡ Let $\mathbf{D}$ be the dielectric displacement in the ether, $\mathbf{H}$ the magnetic force, ρ the volume-density of the charge of an electron, $\mathbf{v}$ the velocity of a point of such a particle, and $\mathbf{F}$ the ponderomotive force, i.e. the force, reckoned per unit charge, which is exerted by the ether on a volume-element of an electron. Then, if we use a fixed system of co-ordinates,

$$\left.\begin{aligned} \operatorname{div} \mathbf{D} &= \rho, \ \operatorname{div} \mathbf{H} = 0, \\ \operatorname{curl} \mathbf{H} &= \frac{1}{c}\left(\frac{\partial \mathbf{D}}{\partial t} + \rho \mathbf{v}\right), \\ \operatorname{curl} \mathbf{D} &= -\frac{1}{c}\frac{\partial \mathbf{H}}{\partial t}, \\ \mathbf{F} &= \mathbf{D} + \frac{1}{c}[\mathbf{v}\,.\,\mathbf{H}]. \end{aligned}\right\} \quad . \quad . \quad . \quad (2)$$

I shall now suppose that the system as a whole moves in the direction of x with a constant velocity v, and I shall denote by $\mathbf{u}$ any velocity which a point of an electron may have in addition to this, so that

$$\mathfrak{v}_x = v + \mathfrak{u}_x, \quad \mathfrak{v}_y = \mathfrak{u}_y, \quad \mathfrak{v}_z = \mathfrak{u}_z.$$

* Poincaré, Rapports du Congrès de physique de 1900, Paris, 1, pp. 22, 23.

† Lorentz, Zittingsverslag Akad. v. Wet., 7, 1899, p. 507; Amsterdam Proc., 1898-99, p. 427.

‡ "M.E.," § 2.

If the equations (2) are at the same time referred to axes moving with the system, they become

$$div\ \mathbf{D} = \rho, \quad div\ \mathbf{H} = 0,$$

$$\frac{\partial \mathrm{H}_z}{\partial y} - \frac{\partial \mathrm{H}_y}{\partial z} = \frac{1}{c}\left(\frac{\partial}{\partial t} - v\frac{\partial}{\partial x}\right)\mathrm{D}_x + \frac{1}{c}\rho(v + u_x),$$

$$\frac{\partial \mathrm{H}_x}{\partial z} - \frac{\partial \mathrm{H}_z}{\partial x} = \frac{1}{c}\left(\frac{\partial}{\partial t} - v\frac{\partial}{\partial x}\right)\mathrm{D}_y + \frac{1}{c}\rho u_y,$$

$$\frac{\partial \mathrm{H}_y}{\partial x} - \frac{\partial \mathrm{H}_x}{\partial y} = \frac{1}{c}\left(\frac{\partial}{\partial t} - v\frac{\partial}{\partial x}\right)\mathrm{D}_z + \frac{1}{c}\rho u_z,$$

$$\frac{\partial \mathrm{D}_z}{\partial y} - \frac{\partial \mathrm{D}_y}{\partial z} = -\frac{1}{c}\left(\frac{\partial}{\partial t} - v\frac{\partial}{\partial x}\right)\mathrm{H}_x,$$

$$\frac{\partial \mathrm{D}_x}{\partial z} - \frac{\partial \mathrm{D}_z}{\partial x} = -\frac{1}{c}\left(\frac{\partial}{\partial t} - v\frac{\partial}{\partial x}\right)\mathrm{H}_y,$$

$$\frac{\partial \mathrm{D}_y}{\partial x} - \frac{\partial \mathrm{D}_x}{\partial y} = -\frac{1}{c}\left(\frac{\partial}{\partial t} - v\frac{\partial}{\partial x}\right)\mathrm{H}_z,$$

$$\mathrm{F}_x = \mathrm{D}_x + \frac{1}{c}(u_y\mathrm{H}_z - u_z\mathrm{H}_y),$$

$$\mathrm{F}_y = \mathrm{D}_y - \frac{1}{c}v\mathrm{H}_z + \frac{1}{c}(u_z\mathrm{H}_x - u_x\mathrm{H}_z),$$

$$\mathrm{F}_z = \mathrm{D}_z + \frac{1}{c}v\mathrm{H}_y + \frac{1}{c}(u_x\mathrm{H}_y - u_y\mathrm{H}_x).$$

§ 4. We shall further transform these formulæ by a change of variables. Putting

$$\frac{c^2}{c^2 - v^2} = \beta^2, \quad . \quad . \quad . \quad . \quad (3)$$

and understanding by l another numerical quantity, to be determined further on, I take as new independent variables

$$x' = \beta l x, \ y' = ly, \ z' = lz, \quad . \quad . \quad . \quad (4)$$

$$t' = \frac{l}{\beta}t - \beta l\frac{v}{c^2}x, \quad . \quad . \quad . \quad (5)$$

and I define two new vectors $\mathbf{D}'$ and $\mathbf{H}'$ by the formulæ

$$D'_x = \frac{1}{l^2}D_x,\ D'_y = \frac{\beta}{l^2}\left(D_y - \frac{v}{c}H_z\right),\ D'_z = \frac{\beta}{l^2}\left(D_z + \frac{v}{c}H_y\right),$$

$$H'_x = \frac{1}{l^2}H_x,\ H'_y = \frac{\beta}{l^2}\left(H_y + \frac{v}{c}D\ \right),\ H'_z = \frac{\beta}{l^2}\left(H_z - \frac{v}{c}D_y\right),$$

for which, on account of (3), we may also write

$$\left.\begin{aligned} D_x &= l^2 D'_x,\ D_y = \beta l^2\left(D'_y + \frac{v}{c}H'_z\right),\ D_z = \beta l^2\left(D'_z - \frac{v}{c}H'_y\right) \\ H_x &= l^2 H'_x,\ H_y = \beta l^2\left(H'_y - \frac{v}{c}D'_z\right),\ H_z = \beta l^2\left(H'_z + \frac{v}{c}D'_y\right) \end{aligned}\right\} \quad (6)$$

As to the coefficient l, it is to be considered as a function of v, whose value is 1 for $v = 0$, and which, for small values of v, differs from unity no more than by a quantity of the second order.

The variable t' may be called the "local time"; indeed, for $\beta = 1$, $l = 1$ it becomes identical with what I formerly denoted by this name.

If, finally, we put

$$\frac{1}{\beta l^3}\rho = \rho' \quad . \quad . \quad . \quad . \quad (7)$$

$$\beta^2 u_x = u'_x,\ \beta u_y = u'_y,\ \beta u_z = u'_z, \quad . \quad . \quad (8)$$

these latter quantities being considered as the components of a new vector $\mathbf{u}'$, the equations take the following form :—

$$\left.\begin{aligned} \text{div}'\,\mathbf{D}' &= \left(1 - \frac{vu'_x}{c^2}\right)\rho',\ \text{div}'\,\mathbf{H}' = 0, \\ \text{curl}'\,\mathbf{H}' &= \frac{1}{c}\left(\frac{\partial \mathbf{D}'}{\partial t'} + \rho'\mathbf{u}'\right), \\ \text{curl}'\,\mathbf{D}' &= -\frac{1}{c}\frac{\partial \mathbf{H}'}{\partial t'}, \end{aligned}\right\} \quad (9)$$

$$\left.\begin{aligned} F_x &= l^2\{D'_x + \frac{1}{c}(u'_y H'_z - u'_z H'_y) + \frac{v}{c^2}(u'_y D'_y + u'_z D'_z)\}, \\ F_y &= \frac{l^2}{\beta}\{D'_y + \frac{1}{c}(u'_z H'_x - u'_x H'_z) - \frac{v}{c^2}u'_x D'_y\}, \\ F_z &= \frac{l^2}{\beta}\{D'_z + \frac{1}{c}(u'_x H'_y - u'_y H'_x) - \frac{v}{c^2}u'_x D'_z\}. \end{aligned}\right\} \quad (10)$$

The meaning of the symbols div′ and curl′ in (9) is similar

to that of div and curl in (2); only, the differentiations with respect to x, y, z are to be replaced by the corresponding ones with respect to x', y', z'.

§ 5. The equations (9) lead to the conclusion that the vectors **D**′ and **H**′ may be represented by means of a scalar potential ϕ' and a vector potential **A**′. These potentials satisfy the equations *

$$\nabla'^2\phi' - \frac{1}{c^2}\frac{\partial^2\phi'}{\partial t'^2} = -\rho' \quad . \quad . \quad . \quad (11)$$

$$\nabla'^2\mathbf{A}' - \frac{1}{c^2}\frac{\partial^2\mathbf{A}'}{\partial t'^2} = -\frac{1}{c}\rho'\mathbf{u}', \quad . \quad . \quad (12)$$

and in terms of them **D**′ and **H**′ are given by

$$\mathbf{D}' = -\frac{1}{c}\frac{\partial\mathbf{A}'}{\partial t'} - \text{grad}'\,\phi' + \frac{v}{c}\text{grad}'\,\mathrm{A}'_x \quad . \quad (13)$$

$$\mathbf{H}' = \text{curl}'\,\mathbf{A}' \quad . \quad . \quad . \quad . \quad (14)$$

The symbol ∇'^2 is an abbreviation for $\frac{\partial^2}{\partial x'^2} + \frac{\partial^2}{\partial y'^2} + \frac{\partial^2}{\partial z'^2}$, and grad′ ϕ' denotes a vector whose components are

$$\frac{\partial\phi'}{\partial x'}, \quad \frac{\partial\phi'}{\partial y'}, \quad \frac{\partial\phi'}{\partial z'}.$$

The expression grad′ A'_x has a similar meaning.

In order to obtain the solution of (11) and (12) in a simple form, we may take x', y', z' as the co-ordinates of a point P′ in a space S′, and ascribe to this point, for each value of t', the values of ρ', **u**′, ϕ', **A**′, belonging to the corresponding point P (x, y, z) of the electromagnetic system. For a definite value t' of the fourth independent variable, the potentials ϕ' and **A**′ at the point P of the system or at the corresponding point P′ of the space S′, are given by †

$$\phi' = \frac{1}{4\pi}\int\frac{[\rho']}{r'}d\mathrm{S}' \quad . \quad . \quad . \quad (15)$$

$$\mathbf{A}' = \frac{1}{4\pi c}\int\frac{[\rho'\mathbf{u}']}{r'}d\mathrm{S}' \quad . \quad . \quad . \quad (16)$$

* "M.E.," §§ 4 and 10. † *Ibid.*, §§ 5 and 10.

Here dS' is an element of the space S', r' its distance from P', and the brackets serve to denote the quantity ρ' and the vector $\rho'\mathbf{u}'$ such as they are in the element dS', for the value $t' - r'/c$ of the fourth independent variable.

Instead of (15) and (16) we may also write, taking into account (4) and (7),

$$\phi' = \frac{1}{4\pi}\int\frac{[\rho]}{r}dS \quad . \quad . \quad . \quad (17)$$

$$\mathbf{A}' = \frac{1}{4\pi c}\int\frac{[\rho\mathbf{u}]}{r}dS, \quad . \quad . \quad . \quad (18)$$

the integrations now extending over the electromagnetic system itself. It should be kept in mind that in these formulæ r' does not denote the distance between the element dS and the point (x, y, z) for which the calculation is to be performed. If the element lies at the point (x_1, y_1, z_1), we must take

$$r' = l\sqrt{\beta^2(x - x_1)^2 + (y - y_1)^2 + (z - z_1)^2}.$$

It is also to be remembered that, if we wish to determine ϕ' and $\mathbf{A}'$ for the instant at which the local time in P is t', we must take ρ and $\rho\mathbf{u}'$, such as they are in the element dS at the instant at which the local time of that element is $t' - r'/c$.

§ 6. It will suffice for our purpose to consider two special cases. The first is that of an electrostatic system, i.e. a system having no other motion but the translation with the velocity v. In this case $\mathbf{u}' = 0$, and therefore, by (12), $\mathbf{A}' = 0$. Also, ϕ' is independent of t', so that the equations (11), (13), and (14) reduce to

$$\left.\begin{aligned} \nabla'^2\phi' &= -\rho', \\ \mathbf{D}' &= -\operatorname{grad}'\phi', \\ \mathbf{H}' &= 0 \end{aligned}\right\} \quad . \quad . \quad . \quad (19)$$

After having determined the vector $\mathbf{D}'$ by means of these equations, we know also the ponderomotive force acting on electrons that belong to the system. For these the formulæ (10) become, since $\mathbf{u}' = 0$,

$$F_x = l^2 D'_x, \ F = \frac{l^2}{\beta}D'_y, \ F_z = \frac{l^2}{\beta}D'_z \quad . \quad . \quad (20)$$

The result may be put in a simple form if we compare the moving system Σ, with which we are concerned, to another electrostatic system Σ' which remains at rest, and into which Σ is changed if the dimensions parallel to the axis of x are multiplied by βl, and the dimensions which have the direction of y or that of z, by l—a deformation for which $(\beta l, l, l)$ is an appropriate symbol. In this new system, which we may suppose to be placed in the above-mentioned space S′, we shall give to the density the value ρ', determined by (7), so that the charges of corresponding elements of volume and of corresponding electrons are the same in Σ and Σ'. Then we shall obtain the forces acting on the electrons of the moving system Σ, if we first determine the corresponding forces in Σ', and next multiply their components in the direction of the axis of x by l^2, and their components perpendicular to that axis by $\frac{l^2}{\beta}$. This is conveniently expressed by the formula

$$\mathbf{F}(\Sigma) = \left(l^2, \frac{l^2}{\beta}, \frac{l^2}{\beta}\right)\mathbf{F}(\Sigma') \quad . \quad . \quad . \quad (21)$$

It is further to be remarked that, after having found $\mathbf{D}'$ by (19), we can easily calculate the electromagnetic momentum in the moving system, or rather its component in the direction of the motion. Indeed, the formula

$$\mathbf{G} = \frac{1}{c}\int[\mathbf{D}\,.\,\mathbf{H}]dS$$

shows that

$$G_x = \frac{1}{c}\int(D_yH_z - D_zH_y)\,dS.$$

Therefore, by (6), since $\mathbf{H}' = 0$

$$G_x = \frac{\beta^2 l^4 v}{c^2}\int(D_y'^2 + D_z'^2)dS = \frac{\beta l v}{c^2}\int(D_y'^2 + D_z'^2)dS'. \quad (22)$$

§ 7. Our second special case is that of a particle having an electric moment, i.e. a small space S, with a total charge $\int\rho dS = 0$, but with such a distribution of density that the integrals $\int \rho x dS$, $\int \rho y dS$, $\int \rho z dS$ have values differing from 0. Let ξ, μ, ζ be the co-ordinates, taken relatively to a fixed point A of the particle, which may be called its centre, and let the electric moment be defined as a vector **P** whose components are

$$P_x = \int \rho\xi dS, \; P_y = \int \rho\eta dS, \; P_z = \int \rho\zeta dS \quad . \quad . \quad (23)$$

Then

$$\frac{dP_x}{dt} = \int \rho u_x dS, \; \frac{dP_y}{dt} = \int \rho u_y dS, \; \frac{dP_z}{dt} = \int \rho u_z dS \; . \quad (24)$$

Of course, if ξ, η, ζ are treated as infinitely small, u_x, u_y, u_z must be so likewise. We shall neglect squares and products of these six quantities.

We shall now apply the equation (17) to the determination of the scalar potential ϕ' for an exterior point P (x, y, z), at a finite distance from the polarized particle, and for the instant at which the local time of this point has some definite value t'. In doing so, we shall give the symbol $[\rho]$, which, in (17), relates to the instarrt at which the local time in dS is $t' - r'/c$, a slightly different meaning. Distinguishing by r'_0 the value of r' for the centre A, we shall understand by $[\rho]$ the value of the density existing in the element dS at the point (ξ, η, ζ), at the instant t_0 at which the local time of A is $t' - r_0/c$.

It may be seen from (5) that this instant precedes that for which we have to take the numerator in (17) by

$$\beta^2\frac{v\xi}{c^2} + \frac{\beta(r'_0 - r')}{lc} = \beta^2\frac{v\xi}{c^2} + \frac{\beta}{lc}\left(\xi\frac{\partial r'}{\partial x} + \eta\frac{\partial r'}{\partial y} + \zeta\frac{\partial r'}{\partial z}\right)$$

units of time. In this last expression we may put for the differential coefficients their values at the point A.

In (17) we have now to replace $[\rho]$ by

$$[\rho] + \beta^2\frac{v\xi}{c^2}\left[\frac{\partial\rho}{\partial t}\right] + \frac{\beta}{lc}\left(\xi\frac{\partial r'}{\partial x} + \eta\frac{\partial r'}{\partial y} + \zeta\frac{\partial r'}{\partial z}\right)\left[\frac{\partial\rho}{\partial t}\right] \qquad (25)$$

where $\left[\frac{\partial\rho}{\partial t}\right]$ relates again to the time t_0. Now, the value of t' for which the calculations are to be performed having been chosen, this time t_0 will be a function of the co-ordinates x, y, z of the exterior point P. The value of $[\rho]$ will therefore depend on these co-ordinates in such a way that

$$\frac{\partial[\rho]}{\partial x} = -\frac{\beta}{lc}\frac{\partial r'}{\partial x}\left[\frac{\partial\rho}{\partial t}\right], \text{ etc.}$$

by which (25) becomes

$$[\rho] + \beta^2\frac{v\xi}{c^2}\left[\frac{\partial\rho}{\partial t}\right] - \left(\xi\frac{\partial[\rho]}{\partial x} + \eta\frac{\partial[\rho]}{\partial y} + \zeta\frac{\partial[\rho]}{\partial z}\right).$$

Again, if henceforth we understand by r' what has above been called r'_0, the factor $\frac{1}{r'}$ must be replaced by

$$\frac{1}{r'} - \xi\frac{\partial}{\partial x}\left(\frac{1}{r'}\right) - \eta\frac{\partial}{\partial y}\left(\frac{1}{r'}\right) - \zeta\frac{\partial}{\partial z}\left(\frac{1}{r'}\right),$$

so that after all, in the integral (17), the element dS is multiplied by

$$\frac{[\rho]}{r'} + \frac{\beta^2 v\xi}{c^2 r'}\left[\frac{\partial\rho}{\partial t}\right] - \frac{\partial}{\partial x}\frac{\xi[\rho]}{r'} - \frac{\partial}{\partial y}\frac{\eta[\rho]}{r'} - \frac{\partial}{\partial z}\frac{\zeta[\rho]}{r'}.$$

This is simpler than the primitive form, because neither r', nor the time for which the quantities enclosed in brackets are to be taken, depend on x, y, z. Using (23) and remembering that $\int\rho d\mathrm{S} = 0$, we get

$$\phi' = \frac{\beta^2 v}{4pc^2 r'}\left[\frac{\partial \mathrm{P}_x}{\partial t}\right] - \frac{1}{4p}\left\{\frac{\partial}{\partial x}\frac{[\mathrm{P}_x]}{r'} + \frac{\partial}{\partial y}\frac{[\mathrm{P}_y]}{r'} + \frac{\partial}{\partial z}\frac{[\mathrm{P}_z]}{r'}\right\},$$

a formula in which all the enclosed quantities are to be taken for the instant at which the local time of the centre of the particle is $t' - r'/c$.

We shall conclude these calculations by introducing a new vector **P**$'$, whose components are

$$\mathrm{P}'_x = \beta l \mathrm{P}_x,\ \mathrm{P}'_y = l\mathrm{P}_y,\ \mathrm{P}'_z = l\mathrm{P}_z, \quad . \quad . \quad (26)$$

passing at the same time to x', y', z', t' as independent variables. The final result is

$$\phi' = \frac{v}{4pc^2 r'}\frac{\partial[\mathrm{P}'_x]}{\partial t'} - \frac{1}{4p}\left\{\frac{\partial}{\partial x'}\frac{[\mathrm{P}'_x]}{r'} + \frac{\partial}{\partial y'}\frac{[\mathrm{P}'_y]}{r'} + \frac{\partial}{\partial z'}\frac{[\mathrm{P}'_z]}{r'}\right\}.$$

As to the formula (18) for the vector potential, its transformation is less complicated, because it contains the infinitely small vector $\mathbf{u}'$. Having regard to (8), (24), (26), and (5), I find

$$\mathbf{A}' = \frac{1}{4\pi c r'} \frac{\partial [\mathbf{P}']}{\partial t'}.$$

The field produced by the polarized particle is now wholly determined. The formula (13) leads to

$$\mathbf{D}' = -\frac{1}{4\pi c^2} \frac{\partial^2}{\partial t'^2} \frac{[\mathbf{P}']}{r'} + \frac{1}{4\pi} \operatorname{grad}' \left\{ \frac{\partial}{\partial x'} \frac{[\mathrm{P}'_x]}{r'} + \frac{\partial}{\partial y'} \frac{[\mathrm{P}'_y]}{r'} + \frac{\partial}{\partial z'} \frac{[\mathrm{P}'_z]}{r'} \right\} \quad (27)$$

and the vector $\mathbf{H}'$ is given by (14). We may further use the equations (20), instead of the original formulæ (10), if we wish to consider the forces exerted by the polarized particle on a similar one placed at some distance. Indeed, in the second particle, as well as in the first, the velocities $\mathbf{u}$ may be held to be infinitely small.

It is to be remarked that the formulæ for a system without translation are implied in what precedes. For such a system the quantities with accents become identical to the corresponding ones without accents; also $\beta = 1$ and $l = 1$. The components of (27) are at the same time those of the electric force which is exerted by one polarized particle on another.

§ 8. Thus far we have used only the fundamental equations without any new assumptions. I shall now suppose *that the electrons, which I take to be spheres of radius R in the state of rest, have their dimensions changed by the effect of a translation, the dimensions in the direction of motion becoming βl times and those in perpendicular directions l times smaller.*

In this deformation, which may be represented by $\left(\frac{1}{\beta l}, \frac{1}{l}, \frac{1}{l}\right)$, each element of volume is understood to preserve its charge.

Our assumption amounts to saying that in an electrostatic system Σ, moving with a velocity v, all electrons are flattened ellipsoids with their smaller axes in the direction of motion. If now, in order to apply the theorem of § 6, we subject the system to the deformation (βl, l, l), we shall have again spherical electrons of radius R. Hence, if we alter the relative position of the centres of the electrons in Σ by applying the deformation (βl, l, l), and if, in the points thus obtained, we place the centres of electrons that remain at rest, we shall get a system, identical to the imaginary system Σ', of which we have spoken in § 6. The forces in this system and those in Σ will bear to each other the relation expressed by (21).

In the second place I shall suppose *that the forces between uncharged particles, as well as those between such particles and electrons, are influenced by a translation in quite the same way as the electric forces in an electrostatic system.* In other terms, whatever be the nature of the particles composing a ponderable body, so long as they do not move relatively to each other, we shall have between the forces acting in a system (Σ') without, and the same system (Σ) with a translation, the relation specified in (21), if, as regards the relative position of the particles, Σ' is got from Σ by the deformation (βl, l, l), or Σ from Σ' by the deformation $\left(\frac{1}{\beta l}, \frac{1}{l}, \frac{1}{l}\right)$.

We see by this that, as soon as the resulting force is zero for a particle in Σ', the same must be true for the corresponding particle in Σ. Consequently, if, neglecting the effects of molecular motion, we suppose each particle of a solid body to be in equilibrium under the action of the attractions and repulsions exerted by its neighbours, and if we take for granted that there is but one configuration of equilibrium, we may draw the conclusion that the system Σ', if the velocity v is imparted to it, will *of itself* change into the system Σ. In other terms, the translation will *produce* the deformation $\left(\frac{1}{\beta l}, \frac{1}{l}, \frac{1}{l}\right)$.

The case of molecular motion will be considered in § 12.

It will easily be seen that the hypothesis which was formerly advanced in connexion with Michelson's experiment, is implied in what has now been said. However, the present hypothesis is more general, because the only limitation imposed on the motion is that its velocity be less than that of light.

§ 9. We are now in a position to calculate the electromagnetic momentum of a single electron. For simplicity's sake I shall suppose the charge e to be uniformly distributed over the surface, so long as the electron remains at rest. Then a distribution of the same kind will exist in the system Σ' with which we are concerned in the last integral of (22). Hence

$$\int (D'^2_y + D'^2_z) dS' = \frac{2}{3}\int D'^2 dS' = \frac{e^2}{6\pi}\int_R^\infty \frac{dr}{r^2} = \frac{e^2}{6\pi R},$$

and

$$G_x = \frac{e^2}{6\pi c^2 R}\beta l v.$$

It must be observed that the product βl is a function of v and that, for reasons of symmetry, the vector $\mathbf{G}$ has the direction of the translation. In general, representing by $\mathbf{v}$ the velocity of this motion, we have the vector equation

$$\mathbf{G} = \frac{e^2}{6\pi c^2 R}\beta l \mathbf{v} \quad . \quad . \quad . \quad . \quad (28)$$

Now, every change in the motion of a system will entail a corresponding change in the electromagnetic momentum and will therefore require a certain force, which is given in direction and magnitude by

$$\mathbf{F} = \frac{d\mathbf{G}}{dt} \quad . \quad . \quad . \quad . \quad (29)$$

Strictly speaking, the formula (28) may only be applied in the case of a uniform rectilinear translation. On account of this circumstance—though (29) is always true—the theory of rapidly varying motions of an electron becomes very complicated, the more so, because the hypothesis of § 8 would imply that the direction and amount of the deformation are continually changing. It is, indeed, hardly probable that the form of the electron will be determined solely by the velocity existing at the moment considered.

Nevertheless, provided the changes in the state of motion

be sufficiently slow, we shall get a satisfactory approximation by using (28) at every instant. The application of (29) to such a *quasi-stationary* translation, as it has been called by Abraham,* is a very simple matter. Let, at a certain instant, $\mathbf{a}_1$ be the acceleration in the direction of the path, and $\mathbf{a}_2$ the acceleration perpendicular to it. Then the force $\mathbf{F}$ will consist of two components, having the directions of these accelerations and which are given by

$$\mathbf{F}_1 = m_1\mathbf{a}_1 \text{ and } \mathbf{F}_2 = m_2\mathbf{a}_2,$$

if

$$m_1 = \frac{e^2}{6\pi c^2 \mathrm{R}} \frac{d(\beta l v)}{dv} \text{ and } m_2 = \frac{e^2}{6\pi c^2 \mathrm{R}} \beta l \quad . \quad . \quad (30)$$

Hence, in phenomena in which there is an acceleration in the direction of motion, the electron behaves as if it had a mass m_1; in those in which the acceleration is normal to the path, as if the mass were m_2. These quantities m_1 and m_2 may therefore properly be called the "longitudinal" and "transverse" electromagnetic masses of the electron. I shall suppose *that there is no other, no "true" or "material" mass.*

Since β and l differ from unity by quantities of the order v^2/c^2, we find for very small velocities

$$m_1 = m_2 = \frac{e^2}{6\pi c^2 \mathrm{R}}.$$

This is the mass with which we are concerned, if there are small vibratory motions of the electrons in a system without translation. If, on the contrary, motions of this kind are going on in a body moving with the velocity v in the direction of the axis of x, we shall have to reckon with the mass m_1, as given by (30), if we consider the vibrations parallel to that axis, and with the mass m_2, if we treat of those that are parallel to OY or OZ. Therefore, in short terms, referring by the index Σ to a moving system and by Σ' to one that remains at rest,

$$m(\Sigma) = \left(\frac{d(\beta l v)}{dv}, \beta l, \beta l\right) m(\Sigma') \quad . \quad . \quad (31)$$

* Abraham, Wied. Ann., 10, 1903, p. 105.

§ 10. We can now proceed to examine the influence of the Earth's motion on optical phenomena in a system of transparent bodies. In discussing this problem we shall fix our attention on the variable electric moments in the particles or "atoms" of the system. To these moments we may apply what has been said in § 7. For the sake of simplicity we shall suppose that, in each particle, the charge is concentrated in a certain number of separate electrons, and that the "elastic" forces that act on one of these, and, conjointly with the electric forces, determine its motion, have their origin within the bounds of the *same* atom.

I shall show that, if we start from any given state of motion in a system without translation, we may deduce from it a corresponding state that can exist in the same system after a translation has been imparted to it, the kind of correspondence being as specified in what follows.

(*a*) Let A'_1, A'_2, A'_3, etc., be the centres of the particles in the system without translation (Σ'); neglecting molecular motions we shall assume these points to remain at rest. The system of points A_1, A_2, A_3, etc., formed by the centres of the particles in the moving system Σ, is obtained from A'_1, A'_2, A'_3, etc., by means of a deformation $\left(\frac{1}{\beta l}, \frac{1}{l}, \frac{1}{l}\right)$. According to what has been said in § 8, the centres will of themselves take these positions A'_1, A'_2, A'_3, etc., if originally, before there was a translation, they occupied the positions A_1, A_2, A_3, etc.

We may conceive any point P' in the space of the system Σ' to be displaced by the above deformation, so that a definite point P of Σ corresponds to it. For two corresponding points P' and P we shall define corresponding instants, the one belonging to P', the other to P, by stating that the true time at the first instant is equal to the local time, as determined by (5) for the point P, at the second instant. By corresponding times for two corresponding *particles* we shall understand times that may be said to correspond, if we fix our attention on the *centres* A' and A of these particles.

(*b*) As regards the interior state of the atoms, we shall assume that the configuration of a particle A in Σ at a certain time may be derived by means of the deformation $\left(\frac{1}{\beta l}, \frac{1}{l}, \frac{1}{l}\right)$ from the configuration of the corresponding particle in Σ', such as it is at the corresponding instant. In so far as this assumption relates to the form of the electrons themselves, it is implied in the first hypothesis of § 8.

Obviously, if we start from a state really existing in the system Σ', we have now completely defined a state of the moving system Σ. The question remains, however, whether this state will likewise be a possible one.

In order to judge of this, we may remark in the first place that the electric moments which we have supposed to exist in the moving system and which we shall denote by **P**, will be certain definite functions of the co-ordinates x, y, z of the centres A of the particles, or, as we shall say, of the co-ordinates of the particles themselves, and of the time t. The equations which express the relations between **P** on one hand and x, y, z, t on the other, may be replaced by other equations containing the vectors **P**$'$ defined by (26) and the quantities x', y', z', t' defined by (4) and (5). Now, by the above assumptions a and b, if in a particle A of the moving system, whose co-ordinates are x, y, z, we find an electric moment **P** at the time t, or at the local time t', the vector **P**$'$ given by (26) will be the moment which exists in the other system at the true time t' in a particle whose co-ordinates are x', y', z'. It appears in this way that the equations between **P**$'$, x', y', z', t' are the same for both systems, the difference being only this, that for the system Σ' without translation these symbols indicate the moment, the co-ordinates, and the true time, whereas their meaning is different for the moving system, **P**$'$, x', y', z', t' being here related to the moment **P**, the co-ordinates x, y, z and the general time t in the manner expressed by (26), (4), and (5).

It has already been stated that the equation (27) applies to both systems. The vector **D**$'$ will therefore be the same in Σ' and Σ, provided we always compare corresponding places and times. However, this vector has not the same meaning in the two cases. In Σ' it represents the electric force, in Σ it is related to this force in the way expressed by (20). We may therefore conclude that the ponderomotive forces acting, in Σ and in Σ', on corresponding particles at corresponding instants, bear to each other the relation determined by (21). In virtue of our assumption (*b*), taken in connexion with the second hypothesis of § 8, the same relation will exist between the "elastic" forces; consequently, the formula (21) may also be regarded as indicating the relation between the total forces, acting on corresponding electrons, at corresponding instants.

It is clear that the state we have supposed to exist in the moving system will really be possible if, in Σ and Σ', the products of the mass m and the acceleration of an electron are to each other in the same relation as the forces, i.e. if

$$m\mathbf{a}(\Sigma) = \left(l^2, \frac{l^2}{\beta}, \frac{l^2}{\beta}\right) m\mathbf{a}(\Sigma') \quad . \quad . \quad . \quad (32)$$

Now, we have for the accelerations

$$\mathbf{a}(\Sigma) = \left(\frac{l}{\beta^3}, \frac{l}{\beta^2}, \frac{l}{\beta^2}\right) \mathbf{a}(\Sigma') \quad . \quad . \quad . \quad (33)$$

as may be deduced from (4) and (5), and combining this with (32), we find for the masses

$$m(\Sigma) = (\beta^3 l, \beta l, \beta l) m(\Sigma').$$

If this is compared with (31), it appears that, whatever be the value of l, the condition is always satisfied, as regards the masses with which we have to reckon when we consider vibrations perpendicular to the translation. The only condition we have to impose on l is therefore

$$\frac{d(\beta l v)}{dv} = \beta^3 l.$$

But, on account of (3),

$$\frac{d(\beta v)}{dv} = \beta^3,$$

so that we must put

$$\frac{dl}{dv} = 0, \; l = \text{const.}$$

The value of the constant must be unity, because we know already that, for $v = 0$, $l = 1$.

We are therefore led to suppose *that the influence of a translation on the dimensions (of the separate electrons and of a ponderable body as a whole) is confined to those that have the direction of the motion, these becoming β times smaller than they are in the state of rest.* If this hypothesis is added to those we have already made, we may be sure that two states, the one in the moving system, the other in the same system while at rest, corresponding as stated above, may both be possible. Moreover, this correspondence is not limited to the electric moments of the particles. In corresponding points that are situated either in the ether between the particles, or in that surrounding the ponderable bodies, we shall find at corresponding times the same vector $\mathbf{D}'$ and, as is easily shown, the same vector $\mathbf{H}'$. We may sum up by saying: If, in the system without translation, there is a state of motion in which, at a definite place, the components of $\mathbf{P}$, $\mathbf{D}$, and $\mathbf{H}$ are certain functions of the time, then the same system after it has been put in motion (and thereby deformed) can be the seat of a state of motion in which, at the corresponding place, the components of $\mathbf{P}'$, $\mathbf{D}'$, and $\mathbf{H}'$ are the same functions of the local time.

There is one point which requires further consideration. The values of the masses m_1 and m_2 having been deduced from the theory of quasi-stationary motion, the question arises, whether we are justified in reckoning with them in the case of the rapid vibrations of light. Now it is found on closer examination that the motion of an electron may be treated as quasi-stationary if it changes very little during the time a light-wave takes to travel over a distance equal to the diameter. This condition is fulfilled in optical phenomena, because the diameter of an electron is extremely small in comparison with the wave-length.

§ 11. It is easily seen that the proposed theory can account for a large number of facts.

Let us take in the first place the case of a system without translation, in some parts of which we have continually $\mathbf{P} = 0$, $\mathbf{D} = 0$, $\mathbf{H} = 0$. Then, in the corresponding state for the moving system, we shall have in corresponding parts (or, as we may say, in the same parts of the deformed system) $\mathbf{P}' = 0$, $\mathbf{D}' = 0$, $\mathbf{H}' = 0$. These equations implying $\mathbf{P} = 0$,

D = 0, **H** = 0, as is seen by (26) and (6), it appears that those parts which are dark while the system is at rest, will remain so after it has been put in motion. It will therefore be impossible to detect an influence of the Earth's motion on any optical experiment, made with a terrestrial source of light, in which the geometrical distribution of light and darkness is observed. Many experiments on interference and diffraction belong to this class.

In the second place, if, in two points of a system, rays of light of the same state of polarization are propagated in the same direction, the ratio between the amplitudes in these points may be shown not to be altered by a translation. The latter remark applies to those experiments in which the intensities in adjacent parts of the field of view are compared.

The above conclusions confirm the results which I formerly obtained by a similar train of reasoning, in which, however, the terms of the second order were neglected. They also contain an explanation of Michelson's negative result, more general than the one previously given, and of a somewhat different form; and they show why Rayleigh and Brace could find no signs of double refraction produced by the motion of the Earth.

As to the experiments of Trouton and Noble, their negative result becomes at once clear, if we admit the hypotheses of § 8. It may be inferred from these and from our last assumption (§ 10) that the only effect of the translation must have been a contraction of the whole system of electrons and other particles constituting the charged condenser and the beam and thread of the torsion-balance. Such a contraction does not give rise to a sensible change of direction.

It need hardly be said that the present theory is put forward with all due reserve. Though it seems to me that it can account for all well-established facts, it leads to some consequences that cannot as yet be put to the test of experiment. One of these is that the result of Michelson's experiment must remain negative, if the interfering rays of light are made to travel through some ponderable transparent body.

Our assumption about the contraction of the electrons

cannot in itself be pronounced to be either plausible or inadmissible. What we know about the nature of electrons is very little, and the only means of pushing our way farther will be to test such hypotheses as I have here made. Of course, there will be difficulties, e.g. as soon as we come to consider the rotation of electrons. Perhaps we shall have to suppose that in those phenomena in which, if there is no translation, spherical electrons rotate about a diameter, the points of the electrons in the moving system will describe elliptic paths, corresponding, in the manner specified in § 10, to the circular paths described in the other case.

§ 12. There remain to be said a few words about molecular motion. We may conceive that bodies in which this has a sensible influence or even predominates, undergo the same deformation as the systems of particles of constant relative position of which alone we have spoken till now. Indeed, in two systems of molecules Σ' and Σ, the first without and the second with a translation, we may imagine molecular motions corresponding to each other in such a way that, if a particle in Σ' has a certain position at a definite instant, a particle in Σ occupies at the corresponding instant the corresponding position. This being assumed, we may use the relation (33) between the accelerations in all those cases in which the velocity of molecular motion is very small as compared with v. In these cases the molecular forces may be taken to be determined by the relative positions, independently of the velocities of molecular motion. If, finally, we suppose these forces to be limited to such small distances that, for particles acting on each other, the difference of local times may be neglected, one of the particles, together with those which lie in its sphere of attraction or repulsion, will form a system which undergoes the often mentioned deformation. In virtue of the second hypothesis of § 8 we may therefore apply to the resulting molecular force acting on a particle, the equation (21). Consequently, the proper relation between the forces and the accelerations will exist in the two cases, if we suppose *that the masses of all particles are influenced by a translation to the same degree as the electromagnetic masses of the electrons.*

§ 13. The values (30), which I have found for the longitudinal and transverse masses of an electron, expressed in terms of its velocity, are not the same as those that had been previously obtained by Abraham. The ground for this difference is to be sought solely in the circumstance that, in his theory, the electrons are treated as spheres of invariable dimensions. Now, as regards the transverse mass, the results of Abraham have been confirmed in a most remarkable way by Kaufmann's measurements of the deflexion of radium-rays in electric and magnetic fields. Therefore, if there is not to be a most serious objection to the theory I have now proposed, it must be possible to show that those measurements agree with my values nearly as well as with those of Abraham.

I shall begin by discussing two of the series of measurements published by Kaufmann* in 1902. From each series he has deduced two quantities η and ζ, the "reduced" electric and magnetic deflexions, which are related as follows to the ratio $\gamma = v/c$:—

$$\gamma = k_1 \frac{\zeta}{\eta}, \qquad \psi(\gamma) = \frac{\eta}{k_2 \zeta^2} \quad . \quad . \quad . \quad (34)$$

Here $\psi(\gamma)$ is such a function, that the transverse mass is given by

$$m_2 = \frac{3}{4} \cdot \frac{e^2}{6\pi c^2 R} \psi(\gamma), \quad . \quad . \quad . \quad (35)$$

whereas k_1 and k_2 are constant in each series.

It appears from the second of the formulæ (30) that my theory leads likewise to an equation of the form (35); only Abraham's function $\psi(\gamma)$ must be replaced by

$$\frac{4}{3}\beta = \frac{4}{3}(1 - \gamma^2)^{-1/2}.$$

Hence, my theory requires that, if we substitute this value for $\psi(\gamma)$ in (34), these equations shall still hold. Of course, in seeking to obtain a good agreement, we shall be justified in giving to k_1 and k_- other values than those of Kaufmann, and in taking for every measurement a proper value of the velocity v, or of the ratio γ. Writing sk_1, $\frac{3}{4} k'_2$

* Kaufmann, Physik. Zeitschr., 4, 1902, p. 55.

and γ' for the new values, we may put (34) in the form

$$\gamma' = sk_1\frac{\zeta}{\eta} \quad . \quad . \quad . \quad (36)$$

and

$$(1 - \gamma'^2)^{-1/2} = \frac{\eta}{k'_2\zeta^2} \quad . \quad . \quad . \quad (37)$$

Kaufmann has tested his equations by choosing for k_1 such a value that, calculating γ and k_2 by means of (34), he obtained values for this latter number which, as well as might be, remained constant in each series. This constancy was the proof of a sufficient agreement.

I have followed a similar method, using, however, some of the numbers calculated by Kaufmann. I have computed for each measurement the value of the expression

$$k'_2 = (1 - \gamma'^2)^{1/2}\psi(\gamma)k_2, \quad . \quad . \quad . \quad (38)$$

that may be got from (37) combined with the second of the equations (34). The values of $\psi(\gamma)$ and k_2 have been taken from Kaufmann's tables, and for γ' I have substituted the value he has found for γ, multiplied by s, the latter coefficient being chosen with a view to obtaining a good constancy of (38). The results are contained in the tables on opposite page, corresponding to the Tables III and IV in Kaufmann's paper.

The constancy of k'_2 is seen to come out no less satisfactorily than that of k_2, the more so as in each case the value of s has been determined by means of only two measurements. The coefficient has been so chosen that for these two observations, which were in Table III the first and the last but one, and in Table IV the first and the last, the values of k'_2 should be proportional to those of k_2.

I shall next consider two series from a later publication by Kaufmann,* which have been calculated by Runge † by means of the method of least squares, the coefficients k_1 and k_2 having been determined in such a way that the values of η, calculated, for each observed ζ, from Kaufmann's equations (34), agree as closely as may be with the observed values of η.

* Kaufmann, Gött. Nachr. Math. phys. Kl., 1903, p. 90.

† Runge, *ibid.*, p. 326.

III. $s = 0{\cdot}933$.

γ.	$\psi(\gamma)$.	k_2.	γ'.	k'_2.
0·851	2·147	1·721	0·794	2·246
0·766	1·86	1·736	0·715	2·258
0·727	1·78	1·725	0·678	2·256
0·6615	1·66	1·727	0·617	2·256
0·6075	1·595	1 655	0·567	2·175

IV. $s = 0{\cdot}954$.

γ.	$\psi(\gamma)$.	k_2.	γ'.	k'_2.
0·963	3·28	8·12	0·919	10·36
0·949	2·86	7·99	0·905	9·70
0·933	2·73	7·46	0·890	9·28
0·883	2·31	8·32	0·842	10·36
0·860	2·195	8·09	0·820	10·15
0·830	2·06	8·13	0·792	10·23
0·801	1·96	8·13	0·764	10·28
0·777	1·89	8·04	0·741	10·20
0·752	1·83	8·02	0·717	10·22
0·732	1·785	7·97	0·698	10·18

I have determined by the same condition, likewise using the method of least squares, the constants a and b in the formula

$$\eta^2 = a\zeta^2 + b\zeta^4,$$

which may be deduced from my equations (36) and (37). Knowing a and b, I find γ for each measurement by means of the relation

$$\gamma = \sqrt{a}\frac{\zeta}{\eta}.$$

For two plates on which Kaufmann had measured the electric and magnetic deflexions, the results are as follows (p. 34), the deflexions being given in centimetres.

I have not found time for calculating the other tables in Kaufmann's paper. As they begin, like the table for Plate 15 (next page) with a rather large negative difference between the values of η which have been deduced from the observations and calculated by Runge, we may expect a satisfactory agreement with my formulæ.

Plate No. 15. $a = 0{\cdot}06489$, $b = 0{\cdot}3039$.

ζ	η					γ	
	Observed.	Calculated by R.	Diff.	Calculated by L.	Diff.	Calculated by R.	L.
0·1495	0·0388	0·0404	− 16	0·0400	− 12	0·987	0·951
0·199	0·0548	0·0550	− 2	0·0552	− 4	0·964	0·918
0·2475	0·0716	0·0710	+ 6	0·0715	+ 1	0·930	0·881
0·296	0·0896	0·0887	+ 9	0·0895	+ 1	0·889	0·842
0 3435	0·1080	0·1081	− 1	0·1090	− 10	0·847	0·803
0·391	0·1290	0·1297	− 7	0·1305	− 15	0·804	0·763
0·437	0·1524	0·1527	− 3	0·1532	− 8	0·763	0·727
0·4 25	0·1788	0·1777	+ 11	0·1777	+ 11	0·724	0·692
0·5265	0·2033	0·2039	− 6	0·2033	0	0·688	0·660

Plate No. 19. $a = 0{\cdot}05867$, $b = 0{\cdot}2591$.

ζ	η					γ	
	Observed.	Calculated by R.	Diff.	Calculated by L.	Diff.	Calculated by R.	L.
0·1495	0·0404	0·0388	+ 16	0·0379	+ 25	0·990	0·954
0·199	0·0529	0·0527	+ 2	0·0522	+ 7	0·969	0·923
0·247	0·0678	0·0675	+ 3	0·0674	+ 4	0·939	0·888
0·296	0·0834	0·0842	− 8	0·0844	− 10	0·902	0·849
0·3435	0·1019	0·1022	− 3	0·1026	− 7	0·862	0·811
0·391	0·1219	0·1222	− 3	0·1226	− 7	0·822	0·773
0·437	0·1429	0·1434	− 5	0·1437	− 8	0·782	0·736
0·4825	0·1660	0·1665	− 5	0·1664	− 4	0·744	0·702
0·5265	0·1916	0·1906	+ 10	0·1902	+ 14	0·709	0·671

ON THE ELECTRODYNAMICS OF MOVING BODIES

BY

A. EINSTEIN

Translated from "Zur Elektrodynamik bewegter Körper," Annalen der Physik, 17, 1905.

ON THE ELECTRODYNAMICS OF MOVING BODIES

By A. EINSTEIN

IT is known that Maxwell's electrodynamics—as usually understood at the present time—when applied to moving bodies, leads to asymmetries which do not appear to be inherent in the phenomena. Take, for example, the reciprocal electrodynamic action of a magnet and a conductor. The observable phenomenon here depends only on the relative motion of the conductor and the magnet, whereas the customary view draws a sharp distinction between the two cases in which either the one or the other of these bodies is in motion. For if the magnet is in motion and the conductor at rest, there arises in the neighbourhood of the magnet an electric field with a certain definite energy, producing a current at the places where parts of the conductor are situated. But if the magnet is stationary and the conductor in motion, no electric field arises in the neighbourhood of the magnet. In the conductor, however, we find an electromotive force, to which in itself there is no corresponding energy, but which gives rise—assuming equality of relative motion in the two cases discussed—to electric currents of the same path and intensity as those produced by the electric forces in the former case.

Examples of this sort, together with the unsuccessful attempts to discover any motion of the earth relatively to the "light medium," suggest that the phenomena of electrodynamics as well as of mechanics possess no properties corresponding to the idea of absolute rest. They suggest rather that, as has already been shown to the first order of small quantities, the same laws of electrodynamics and optics will be valid for all frames of reference for which the equations of mechanics hold good.* We will raise this conjecture (the purport of which will hereafter be called the "Principle of Relativity") to the status of a postulate, and also introduce another postulate, which is only apparently irreconcilable with the former, namely, that light is always propagated in empty space with a definite velocity c which is independent of the state of motion of the emitting body. These two postulates suffice for the attainment of a simple and consistent theory of the electrodynamics of moving bodies based on Maxwell's theory for stationary bodies. The introduction of a "luminiferous ether" will prove to be superfluous inasmuch as the view here to be developed will not require an "absolutely stationary space" provided with special properties, nor assign a velocity-vector to a point of the empty space in which electromagnetic processes take place.

The theory to be developed is based—like all electrodynamics—on the kinematics of the rigid body, since the assertions of any such theory have to do with the relationships between rigid bodies (systems of co-ordinates), clocks, and electromagnetic processes. Insufficient consideration of this circumstance lies at the root of the difficulties which the electrodynamics of moving bodies at present encounters.

I. Kinematical Part

§ 1. Definition of Simultaneity

Let us take a system of co-ordinates in which the equations of Newtonian mechanics hold good.† In order to render our presentation more precise and to distinguish this system of co-ordinates verbally from others which will be introduced hereafter, we call it the "stationary system."

If a material point is at rest relatively to this system of co-ordinates, its position can be defined relatively thereto by the employment of rigid standards of measurement and the methods of Euclidean geometry, and can be expressed in Cartesian co-ordinates.

If we wish to describe the *motion* of a material point, we

* The preceding memoir by Lorentz was not at this time known to the author.

† i.e. to the first approximation.

give the values of its co-ordinates as functions of the time. Now we must bear carefully in mind that a mathematical description of this kind has no physical meaning unless we are quite clear as to what we understand by "time." We have to take into account that all our judgments in which time plays a part are always judgments of *simultaneous events.* If, for instance, I say, "That train arrives here at 7 o'clock," I mean something like this: "The pointing of the small hand of my watch to 7 and the arrival of the train are simultaneous events." *

It might appear possible to overcome all the difficulties attending the definition of "time" by substituting "the position of the small hand of my watch" for "time." And in fact such a definition is satisfactory when we are concerned with defining a time exclusively for the place where the watch is located; but it is no longer satisfactory when we have to connect in time series of events occurring at different places, or—what comes to the same thing—to evaluate the times of events occurring at places remote from the watch.

We might, of course, content ourselves with time values determined by an observer stationed together with the watch at the origin of the co-ordinates, and co-ordinating the corresponding positions of the hands with light signals, given out by every event to be timed, and reaching him through empty space. But this co-ordination has the disadvantage that it is not independent of the standpoint of the observer with the watch or clock, as we know from experience. We arrive at a much more practical determination along the following line of thought.

If at the point A of space there is a clock, an observer at A can determine the time values of events in the immediate proximity of A by finding the positions of the hands which are simultaneous with these events. If there is at the point B of space another clock in all respects resembling the one at A, it is possible for an observer at B to determine the time values of events in the immediate neighbourhood of B. But it is not possible without further assumption to compare, in

* We shall not here discuss the inexactitude which lurks in the concept of simultaneity of two events at approximately the same place, which can only be removed by an abstraction.

respect of time, an event at A with an event at B. We have so far defined only an " A time " and a " B time." We have not defined a common " time " for A and B, for the latter cannot be defined at all unless we establish *by definition* that the " time " required by light to travel from A to B equals the " time " it requires to travel from B to A. Let a ray of light start at the " A time " t_A from A towards B, let it at the " B time" t_B be reflected at B in the direction of A, and arrive again at A at the " A time " t'_A

In accordance with definition the two clocks synchronize if

$$t_B - t_A = t'_A - t_B.$$

We assume that this definition of synchronism is free from contradictions, and possible for any number of points; and that the following relations are universally valid :—

1. If the clock at B synchronizes with the clock at A, the clock at A synchronizes with the clock at B.

2. If the clock at A synchronizes with the clock at B and also with the clock at C, the clocks at B and C also synchronize with each other.

Thus with the help of certain imaginary physical experiments we have settled what is to be understood by synchronous stationary clocks located at different places, and have evidently obtained a definition of " simultaneous," or " synchronous," and of " time." The " time " of an event is that which is given simultaneously with the event by a stationary clock located at the place of the event, this clock being synchronous, and indeed synchronous for all time determinations, with a specified stationary clock.

In agreement with experience we further assume the quantity

$$\frac{2AB}{t'_A - t_A} = c,$$

to be a universal constant—the velocity of light in empty space.

It is essential to have time defined by means of stationary clocks in the stationary system, and the time now defined being appropriate to the stationary system we call it " the time of the stationary system."

§ 2. On the Relativity of Lengths and Times

The following reflexions are based on the principle of relativity and on the principle of the constancy of the velocity of light. These two principles we define as follows:—

1. The laws by which the states of physical systems undergo change are not affected, whether these changes of state be referred to the one or the other of two systems of co-ordinates in uniform translatory motion.

2. Any ray of light moves in the "stationary" system of co-ordinates with the determined velocity c, whether the ray be emitted by a stationary or by a moving body. Hence

$$\text{velocity} = \frac{\text{light path}}{\text{time interval}}$$

where time interval is to be taken in the sense of the definition in § 1.

Let there be given a stationary rigid rod; and let its length be l as measured by a measuring-rod which is also stationary. We now imagine the axis of the rod lying along the axis of x of the stationary system of co-ordinates, and that a uniform motion of parallel translation with velocity v along the axis of x in the direction of increasing x is then imparted to the rod. We now inquire as to the length of the moving rod, and imagine its length to be ascertained by the following two operations:—

(*a*) The observer moves together with the given measuring-rod and the rod to be measured, and measures the length of the rod directly by superposing the measuring-rod, in just the same way as if all three were at rest.

(*b*) By means of stationary clocks set up in the stationary system and synchronizing in accordance with § 1, the observer ascertains at what points of the stationary system the two ends of the rod to be measured are located at a definite time. The distance between these two points, measured by the measuring-rod already employed, which in this case is at rest, is also a length which may be designated "the length of the rod."

In accordance with the principle of relativity the length

to be discovered by the operation (*a*)—we will call it "the length of the rod in the moving system"—must be equal to the length l of the stationary rod.

The length to be discovered by the operation (*b*) we will call "the length of the (moving) rod in the stationary system." This we shall determine on the basis of our two principles, and we shall find that it differs from l.

Current kinematics tacitly assumes that the lengths determined by these two operations are precisely equal, or in other words, that a moving rigid body at the epoch t may in geometrical respects be perfectly represented by *the same* body *at rest* in a definite position.

We imagine further that at the two ends A and B of the rod, clocks are placed which synchronize with the clocks of the stationary system, that is to say that their indications correspond at any instant to the "time of the stationary system" at the places where they happen to be. These clocks are therefore "synchronous in the stationary system."

We imagine further that with each clock there is a moving observer, and that these observers apply to both clocks the criterion established in § 1 for the synchronization of two clocks. Let a ray of light depart from A at the time * t_A, let it be reflected at B at the time t_B, and reach A again at the time t'_A. Taking into consideration the principle of the constancy of the velocity of light we find that

$$t_B - t_A = \frac{r_{AB}}{c - v} \text{ and } t'_A - t_B = \frac{r_{AB}}{c + v}$$

where r_{AB} denotes the length of the moving rod—measured in the stationary system. Observers moving with the moving rod would thus find that the two clocks were not synchronous, while observers in the stationary system would declare the clocks to be synchronous.

So we see that we cannot attach any *absolute* signification to the concept of simultaneity, but that two events which, viewed from a system of co-ordinates, are simultaneous, can no longer be looked upon as simultaneous events when en-

* "Time" here denotes "time of the stationary system" and also "position of hands of the moving clock situated at the place under discussion."

visaged from a system which is in motion relatively to that system.

§ 3. Theory of the Transformation of Co-ordinates and Times from a Stationary System to another System in Uniform Motion of Translation Relatively to the Former

Let us in "stationary" space take two systems of co-ordinates, i.e. two systems, each of three rigid material lines, perpendicular to one another, and issuing from a point. Let the axes of X of the two systems coincide, and their axes of Y and Z respectively be parallel. Let each system be provided with a rigid measuring-rod and a number of clocks, and let the two measuring-rods, and likewise all the clocks of the two systems, be in all respects alike.

Now to the origin of one of the two systems (k) let a constant velocity v be imparted in the direction of the increasing x of the other stationary system (K), and let this velocity be communicated to the axes of the co-ordinates, the relevant measuring-rod, and the clocks. To any time of the stationary system K there then will correspond a definite position of the axes of the moving system, and from reasons of symmetry we are entitled to assume that the motion of k may be such that the axes of the moving system are at the time t (this "t" always denotes a time of the stationary system) parallel to the axes of the stationary system.

We now imagine space to be measured from the stationary system K by means of the stationary measuring-rod, and also from the moving system k by means of the measuring-rod moving with it; and that we thus obtain the co-ordinates x, y, z, and ξ, η, ζ respectively. Further, let the time t of the stationary system be determined for all points thereof at which there are clocks by means of light signals in the manner indicated in § 1; similarly let the time τ of the moving system be determined for all points of the moving system at which there are clocks at rest relatively to that system by applying the method, given in § 1, of light signals between the points at which the latter clocks are located.

To any system of values x, y, z, t, which completely defines the place and time of an event in the stationary system, there belongs a system of values ξ, η, ζ, τ, determining that event relatively to the system k, and our task is now to find the system of equations connecting these quantities.

In the first place it is clear that the equations must be *linear* on account of the properties of homogeneity which we attribute to space and time.

If we place $x' = x - vt$, it is clear that a point at rest in the system k must have a system of values x', y, z, independent of time. We first define τ as a function of x', y, z, and t. To do this we have to express in equations that τ is nothing else than the summary of the data of clocks at rest in system k, which have been synchronized according to the rule given in § 1.

From the origin of system k let a ray be emitted at the time τ_0 along the X-axis to x', and at the time τ_1 be reflected thence to the origin of the co-ordinates, arriving there at the time τ_2; we then must have $\frac{1}{2}(\tau_0 + \tau_2) = \tau_1$, or, by inserting the arguments of the function τ and applying the principle of the constancy of the velocity of light in the stationary system:—

$$\tfrac{1}{2}\left[\tau(0,0,0,t)+\tau\left(0,0,0,t+\frac{x'}{c-v}+\frac{x'}{c+v}\right)\right]=\tau\left(x',0,0,t+\frac{x'}{c-v}\right).$$

Hence, if x' be chosen infinitesimally small,

$$\tfrac{1}{2}\left(\frac{1}{c-v}+\frac{1}{c+v}\right)\frac{\partial\tau}{\partial t}=\frac{\partial\tau}{\partial x'}+\frac{1}{c-v}\frac{\partial\tau}{\partial t},$$

or

$$\frac{\partial\tau}{\partial x'}+\frac{v}{c^2-v^2}\frac{\partial\tau}{\partial t}=0.$$

It is to be noted that instead of the origin of the co-ordinates we might have chosen any other point for the point of origin of the ray, and the equation just obtained is therefore valid for all values of x', y, z.

An analogous consideration—applied to the axes of Y and Z—it being borne in mind that light is always propagated along these axes, when viewed from the stationary system, with the velocity $\sqrt{(c^2 - v^2)}$, gives us

$$\frac{\partial\tau}{\partial y}=0,\ \frac{\partial\tau}{\partial z}=0.$$

Since τ is a *linear* function, it follows from these equations that

$$\tau = a\left(t - \frac{v}{c^2 - v^2}x'\right)$$

where a is a function $\phi(v)$ at present unknown, and where for brevity it is assumed that at the origin of k, $\tau = 0$, when $t = 0$.

With the help of this result we easily determine the quantities ξ, η, ζ by expressing in equations that light (as required by the principle of the constancy of the velocity of light, in combination with the principle of relativity) is also propagated with velocity c when measured in the moving system. For a ray of light emitted at the time $\tau = 0$ in the direction of the increasing ξ

$$\xi = c\tau \text{ or } \xi = ac\left(t - \frac{v}{c^2 - v^2}x'\right).$$

But the ray moves relatively to the initial point of k, when measured in the stationary system, with the velocity $c - v$, so that

$$\frac{x'}{c - v} = t.$$

If we insert this value of t in the equation for ξ, we obtain

$$\xi = a\frac{c^2}{c^2 - v^2}x'.$$

In an analogous manner we find, by considering rays moving along the two other axes, that

$$\eta = c\tau = ac\left(t - \frac{v}{c^2 - v^2}x'\right)$$

when

$$\frac{y}{\sqrt{(c^2 - v^2)}} = t,\ x' = 0.$$

Thus

$$\eta = a\frac{c}{\sqrt{(c^2 - v^2)}}y \text{ and } \zeta = a\frac{c}{\sqrt{(c^2 - v^2)}}z.$$

Substituting for x' its value, we obtain

$$\tau = \phi(v)\beta(t - vx/c^2),$$
$$\xi = \phi(v)\beta(x - vt),$$
$$\eta = \phi(v)y,$$
$$\zeta = \phi(v)z,$$

where

$$\beta = \frac{1}{\sqrt{(1 - v^2/c^2)}},$$

and ϕ is an as yet unknown function of v. If no assumption whatever be made as to the initial position of the moving system and as to the zero point of τ, an additive constant is to be placed on the right side of each of these equations.

We now have to prove that any ray of light, measured in the moving system, is propagated with the velocity c, if, as we have assumed, this is the case in the stationary system; for we have not as yet furnished the proof that the principle of the constancy of the velocity of light is compatible with the principle of relativity.

At the time $t = \tau = 0$, when the origin of the co-ordinates is common to the two systems, let a spherical wave be emitted therefrom, and be propagated with the velocity c in system K. If (x, y, z) be a point just attained by this wave, then

$$x^2 + y^2 + z^2 = c^2t^2.$$

Transforming this equation with the aid of our equations of transformation we obtain after a simple calculation

$$\xi^2 + \eta^2 + \zeta^2 = c^2\tau^2.$$

The wave under consideration is therefore no less a spherical wave with velocity of propagation c when viewed in the moving system. This shows that our two fundamental principles are compatible.*

In the equations of transformation which have been developed there enters an unknown function ϕ of v, which we will now determine.

For this purpose we introduce a third system of co-ordin-

* The equations of the Lorentz transformation may be more simply deduced directly from the condition that in virtue of those equations the relation $x^2 + y^2 + z^2 = c^2t^2$ shall have as its consequence the second relation $\xi^2 + \eta^2 + \zeta^2 = c^2\tau^2$.

ates K′, which relatively to the system k is in a state of parallel translatory motion parallel to the axis of X, such that the origin of co-ordinates of system k moves with velocity $-v$ on the axis of X. At the time $t = 0$ let all three origins coincide, and when $t = x = y = z = 0$ let the time t' of the system K′ be zero. We call the co-ordinates, measured in the system K′, x', y', z', and by a twofold application of our equations of transformation we obtain

$$\begin{aligned} t' &= \phi(-v)\beta(-v)(\tau + v\xi/c^2) = \phi(v)\phi(-v)t, \\ x' &= \phi(-v)\beta(-v)(\xi + v\tau) \quad = \phi(v)\phi(-v)x, \\ y' &= \phi(-v)\eta \qquad\qquad\qquad\quad = \phi(v)\phi(-v)y, \\ z' &= \phi(-v)\zeta \qquad\qquad\qquad\quad = \phi(v)\phi(-v)z. \end{aligned}$$

Since the relations between x', y', z' and x, y, z do not contain the time t, the systems K and K′ are at rest with respect to one another, and it is clear that the transformation from K to K′ must be the identical transformation. Thus

$$\phi(v)\phi(-v) = 1.$$

We now inquire into the signification of $\phi(v)$. We give our attention to that part of the axis of Y of system k which lies between $\xi = 0$, $\eta = 0$, $\zeta = 0$ and $\xi = 0$, $\eta = l$, $\zeta = 0$. This part of the axis of Y is a rod moving perpendicularly to its axis with velocity v relatively to system K. Its ends possess in K the co-ordinates

$$x_1 = vt,\ y_1 = \frac{l}{\phi(v)},\ z_1 = 0$$

and

$$x_2 = vt,\ y_2 = 0,\ z_2 = 0.$$

The length of the rod measured in K is therefore $l/\phi(v)$; and this gives us the meaning of the function $\phi(v)$. From reasons of symmetry it is now evident that the length of a given rod moving perpendicularly to its axis, measured in the stationary system, must depend only on the velocity and not on the direction and the sense of the motion. The length of the moving rod measured in the stationary system does not change, therefore, if v and $-v$ are interchanged. Hence follows that $l/\phi(v) = l/\phi(-v)$, or

$$\phi(v) = \phi(-v).$$

It follows from this relation and the one previously found

that $\phi(v) = 1$, so that the transformation equations which have been found become

$$\begin{aligned}\tau &= \beta(t - vx/c^2),\\ \xi &= \beta(x - vt),\\ \eta &= y,\\ \zeta &= z,\end{aligned}$$

where

$$\beta = 1/\sqrt{(1 - v^2/c^2)}.$$

§ 4. Physical Meaning of the Equations Obtained in Respect to Moving Rigid Bodies and Moving Clocks

We envisage a rigid sphere* of radius R, at rest relatively to the moving system k, and with its centre at the origin of co-ordinates of k. The equation of the surface of this sphere moving relatively to the system K with velocity v is

$$\xi^2 + \eta^2 + \zeta^2 = \mathrm{R}^2.$$

The equation of this surface expressed in x, y, z at the time $t = 0$ is

$$\frac{x^2}{(\sqrt{(1 - v^2/c^2)})^2} + y^2 + z^2 = \mathrm{R}^2.$$

A rigid body which, measured in a state of rest, has the form of a sphere, therefore has in a state of motion—viewed from the stationary system—the form of an ellipsoid of revolution with the axes

$$\mathrm{R}\sqrt{(1 - v^2/c^2)},\ \mathrm{R},\ \mathrm{R}.$$

Thus, whereas the Y and Z dimensions of the sphere (and therefore of every rigid body of no matter what form) do not appear modified by the motion, the X dimension appears shortened in the ratio $1 : \sqrt{(1 - v^2/c^2)}$, i.e. the greater the value of v, the greater the shortening. For $v = c$ all moving objects—viewed from the "stationary" system—shrivel up into plain figures. For velocities greater than that of light our deliberations become meaningless; we shall, however, find in what follows, that the velocity of light in our theory plays the part, physically, of an infinitely great velocity.

* That is, a body possessing spherical form when examined at rest.

It is clear that the same results hold good of bodies at rest in the "stationary" system, viewed from a system in uniform motion.

Further, we imagine one of the clocks which are qualified to mark the time t when at rest relatively to the stationary system, and the time τ when at rest relatively to the moving system, to be located at the origin of the co-ordinates of k, and so adjusted that it marks the time τ. What is the rate of this clock, when viewed from the stationary system?

Between the quantities x, t, and τ, which refer to the position of the clock, we have, evidently, $x = vt$ and

$$\tau = \frac{1}{\sqrt{(1 - v^2/c^2)}}(t - vx/c^2).$$

Therefore,

$$\tau = t\sqrt{(1 - v^2/c^2)} = t - (1 - \sqrt{(1 - v^2/c^2)})t$$

whence it follows that the time marked by the clock (viewed in the stationary system) is slow by $1 - \sqrt{(1 - v^2/c^2)}$ seconds per second, or—neglecting magnitudes of fourth and higher order—by $\frac{1}{2}v^2/c^2$.

From this there ensues the following peculiar consequence. If at the points A and B of K there are stationary clocks which, viewed in the stationary system, are synchronous; and if the clock at A is moved with the velocity v along the line AB to B, then on its arrival at B the two clocks no longer synchronize, but the clock moved from A to B lags behind the other which has remained at B by $\frac{1}{2}tv^2/c^2$ (up to magnitudes of fourth and higher order), t being the time occupied in the journey from A to B.

It is at once apparent that this result still holds good if the clock moves from A to B in any polygonal line, and also when the points A and B coincide.

If we assume that the result proved for a polygonal line is also valid for a continuously curved line, we arrive at this result: If one of two synchronous clocks at A is moved in a closed curve with constant velocity until it returns to A, the journey lasting t seconds, then by the clock which has remained at rest the travelled clock on its arrival at A will be $\frac{1}{2}tv^2/c^2$ second slow. Thence we conclude that a balance-clock * at the equator must go more slowly, by a very small amount, than a precisely similar clock situated at one of the poles under otherwise identical conditions.

§ 5. The Composition of Velocities

In the system k moving along the axis of X of the system K with velocity v, let a point move in accordance with the equations

$$\xi = w_\xi \tau,\ \eta = w_\eta \tau,\ \zeta = 0,$$

where w_ξ and w_η denote constants.

Required: the motion of the point relatively to the system K. If with the help of the equations of transformation developed in § 3 we introduce the quantities x, y, z, t into the equations of motion of the point, we obtain

$$x = \frac{w_\xi + v}{1 + vw_\xi/c^2}t,$$

$$y = \frac{\sqrt{(1 - v^2/c^2)}}{1 + vw_\xi/c^2}w_\eta t,$$

$$z = 0.$$

Thus the law of the parallelogram of velocities is valid according to our theory only to a first approximation. We set

$$V^2 = \left(\frac{dx}{dt}\right)^2 + \left(\frac{dy}{dt}\right)^2,$$

$$w^2 = w_\xi^2 + w_\eta^2,$$

$$a = \tan^{-1} w_y/w_x,$$

a is then to be looked upon as the angle between the velocities v and w. After a simple calculation we obtain

$$V = \frac{\sqrt{[(v^2 + w^2 + 2vw \cos a) - (vw \sin a/c^2)^2]}}{1 + vw \cos a/c^2}.$$

It is worthy of remark that v and w enter into the expression for the resultant velocity in a symmetrical manner. If w also has the direction of the axis of X, we get

$$V = \frac{v + w}{1 + vw/c^2},$$

* Not a pendulum-clock, which is physically a system to which the Earth belongs. This case had to be excluded.

It follows from this equation that from a composition of two velocities which are less than c, there always results a velocity less than c. For if we set $v = c - \kappa$, $w = c - \lambda$, κ and λ being positive and less than c, then

$$V = c\frac{2c - \kappa - \lambda}{2c - \kappa - \lambda + \kappa\lambda/c} < c.$$

It follows, further, that the velocity of light c cannot be altered by composition with a velocity less than that of light. For this case we obtain

$$V = \frac{c + w}{1 + w/c} = c.$$

We might also have obtained the formula for V, for the case when v and w have the same direction, by compounding two transformations in accordance with § 3. If in addition to the systems K and k figuring in § 3 we introduce still another system of co-ordinates k' moving parallel to k, its initial point moving on the axis of X with the velocity w, we obtain equations between the quantities x, y, z, t and the corresponding quantities of k', which differ from the equations found in § 3 only in that the place of "v" is taken by the quantity

$$\frac{v + w}{1 + vw/c^2};$$

from which we see that such parallel transformations—necessarily—form a group.

We have now deduced the requisite laws of the theory of kinematics corresponding to our two principles, and we proceed to show their application to electrodynamics.

II. Electrodynamical Part

§ 6. Transformation of the Maxwell-Hertz Equations for Empty Space. On the Nature of the Electromotive Forces Occurring in a Magnetic Field During Motion

Let the Maxwell-Hertz equations for empty space hold good for the stationary system K, so that we have

$$\frac{1}{c}\frac{\partial \mathrm{X}}{\partial t}=\frac{\partial \mathrm{N}}{\partial y}-\frac{\partial \mathrm{M}}{\partial z},\quad \frac{1}{c}\frac{\partial \mathrm{L}}{\partial t}=\frac{\partial \mathrm{Y}}{\partial z}-\frac{\partial \mathrm{Z}}{\partial y},$$

$$\frac{1}{c}\frac{\partial \mathrm{Y}}{\partial t}=\frac{\partial \mathrm{L}}{\partial z}-\frac{\partial \mathrm{N}}{\partial x},\quad \frac{1}{c}\frac{\partial \mathrm{M}}{\partial t}=\frac{\partial \mathrm{Z}}{\partial x}-\frac{\partial \mathrm{X}}{\partial z},$$

$$\frac{1}{c}\frac{\partial \mathrm{Z}}{\partial t}=\frac{\partial \mathrm{M}}{\partial x}-\frac{\partial \mathrm{L}}{\partial y},\quad \frac{1}{c}\frac{\partial \mathrm{N}}{\partial t}=\frac{\partial \mathrm{X}}{\partial y}-\frac{\partial \mathrm{Y}}{\partial x},$$

where (X, Y, Z) denotes the vector of the electric force, and (L, M, N) that of the magnetic force.

If we apply to these equations the transformation developed in § 3, by referring the electromagnetic processes to the system of co-ordinates there introduced, moving with the velocity v, we obtain the equations

$$\frac{1}{c}\frac{\partial \mathrm{X}}{\partial \tau}=\frac{\partial}{\partial \eta}\left\{\beta\left(\mathrm{N}-\frac{v}{c}\mathrm{Y}\right)\right\}-\frac{\partial}{\partial \zeta}\left\{\beta\left(\mathrm{M}+\frac{v}{c}\mathrm{Z}\right)\right\},$$

$$\frac{1}{c}\frac{\partial}{\partial \tau}\left\{\beta\left(\mathrm{Y}-\frac{v}{c}\mathrm{N}\right)\right\}=\frac{\partial \mathrm{L}}{\partial \xi}-\frac{\partial}{\partial \zeta}\left\{\beta\left(\mathrm{N}-\frac{v}{c}\mathrm{Y}\right)\right\}.$$

$$\frac{1}{c}\frac{\partial}{\partial \tau}\left\{\beta\left(\mathrm{Z}+\frac{v}{c}\mathrm{M}\right)\right\}=\frac{\partial}{\partial \xi}\left\{\beta\left(\mathrm{M}+\frac{v}{c}\mathrm{Z}\right)\right\}-\frac{\partial \mathrm{L}}{\partial \eta},$$

$$\frac{1}{c}\frac{\partial \mathrm{L}}{\partial \tau}=\frac{\partial}{\partial \zeta}\left\{\beta\left(\mathrm{Y}-\frac{v}{c}\mathrm{N}\right)\right\}-\frac{\partial}{\partial \eta}\left\{\beta\left(\mathrm{Z}+\frac{v}{c}\mathrm{M}\right)\right\},$$

$$\frac{1}{c}\frac{\partial}{\partial \tau}\left\{\beta\left(\mathrm{M}+\frac{v}{c}\mathrm{Z}\right)\right\}=\frac{\partial}{\partial \xi}\left\{\beta\left(\mathrm{Z}+\frac{v}{c}\mathrm{M}\right)\right\}-\frac{\partial \mathrm{X}}{\partial \zeta},$$

$$\frac{1}{c}\frac{\partial}{\partial \tau}\left\{\beta\left(\mathrm{N}-\frac{v}{c}\mathrm{Y}\right)\right\}=\frac{\partial \mathrm{X}}{\partial \eta}-\frac{\partial}{\partial \xi}\left\{\beta\left(\mathrm{Y}-\frac{v}{c}\mathrm{N}\right)\right\},$$

where

$$\beta=1/\sqrt{(1-v^2/c^2)}.$$

Now the principle of relativity requires that if the Maxwell-Hertz equations for empty space hold good in system K, they also hold good in system k; that is to say that the vectors of the electric and the magnetic force—(X′, Y′, Z′) and (L′, M′, N′)—of the moving system k, which are defined by their ponderomotive effects on electric or magnetic masses respectively, satisfy the following equations :—

$$\frac{1}{c}\frac{\partial X'}{\partial \tau} = \frac{\partial N'}{\partial \eta} - \frac{\partial M'}{\partial \zeta}, \quad \frac{1}{c}\frac{\partial L'}{\partial \tau} = \frac{\partial Y'}{\partial \zeta} - \frac{\partial Z'}{\partial \eta},$$

$$\frac{1}{c}\frac{\partial Y'}{\partial \tau} = \frac{\partial L'}{\partial \zeta} - \frac{\partial N'}{\partial \xi}, \quad \frac{1}{c}\frac{\partial M'}{\partial \tau} = \frac{\partial Z'}{\partial \xi} - \frac{\partial X'}{\partial \zeta},$$

$$\frac{1}{c}\frac{\partial Z'}{\partial \tau} = \frac{\partial M'}{\partial \xi} - \frac{\partial L'}{\partial \eta}, \quad \frac{1}{c}\frac{\partial N'}{\partial \tau} = \frac{\partial X'}{\partial \eta} - \frac{\partial Y'}{\partial \xi}.$$

Evidently the two systems of equations found for system k must express exactly the same thing, since both systems of equations are equivalent to the Maxwell-Hertz equations for system K. Since, further, the equations of the two systems agree, with the exception of the symbols for the vectors, it follows that the functions occurring in the systems of equations at corresponding places must agree, with the exception of a factor $\psi(v)$, which is common for all functions of the one system of equations, and is independent of ξ, η, ζ and τ but depends upon v. Thus we have the relations

$$X' = \psi(v)X, \qquad L' = \psi(v)L,$$

$$Y' = \psi(v)\beta\left(Y - \frac{v}{c}N\right), \quad M' = \psi(v)\beta\left(M + \frac{v}{c}Z\right),$$

$$Z' = \psi(v)\beta\left(Z + \frac{v}{c}M\right), \quad N' = \psi(v)\beta\left(N - \frac{v}{c}Y\right).$$

If we now form the reciprocal of this system of equations, firstly by solving the equations just obtained, and secondly by applying the equations to the inverse transformation (from k to K), which is characterized by the velocity $-v$, it follows, when we consider that the two systems of equations thus obtained must be identical, that $\psi(v)\psi(-v) = 1$. Further, from reasons of symmetry * $\psi(v) = \psi(-v)$, and therefore

$$\psi(v) = 1,$$

and our equations assume the form

* If, for example, $X = Y = Z = L = M = 0$, and $N \neq 0$, then from reasons of symmetry it is clear that when v changes sign without changing its numerical value, Y' must also change sign without changing its numerical value.

$$X' = X, \qquad L' = L,$$

$$Y' = \beta\left(Y - \frac{v}{c}N\right), \quad M' = \beta\left(M + \frac{v}{c}Z\right),$$

$$Z' = \beta\left(Z + \frac{v}{c}M\right), \quad N' = \beta\left(N - \frac{v}{c}Y\right).$$

As to the interpretation of these equations we make the following remarks: Let a point charge of electricity have the magnitude "one" when measured in the stationary system K, i.e. let it when at rest in the stationary system exert a force of one dyne upon an equal quantity of electricity at a distance of one cm. By the principle of relativity this electric charge is also of the magnitude "one" when measured in the moving system. If this quantity of electricity is at rest relatively to the stationary system, then by definition the vector (X, Y, Z) is equal to the force acting upon it. If the quantity of electricity is at rest relatively to the moving system (at least at the relevant instant), then the force acting upon it, measured in the moving system, is equal to the vector (X′, Y′, Z′). Consequently the first three equations above allow themselves to be clothed in words in the two following ways:—

1. If a unit electric point charge is in motion in an electromagnetic field, there acts upon it, in addition to the electric force, an "electromotive force" which, if we neglect the terms multiplied by the second and higher powers of v/c, is equal to the vector-product of the velocity of the charge and the magnetic force, divided by the velocity of light. (Old manner of expression.)

2. If a unit electric point charge is in motion in an electromagnetic field, the force acting upon it is equal to the electric force which is present at the locality of the charge, and which we ascertain by transformation of the field to a system of co-ordinates at rest relatively to the electrical charge. (New manner of expression.)

The analogy holds with "magnetomotive forces." We see that electromotive force plays in the developed theory merely the part of an auxiliary concept, which owes its introduction to the circumstance that electric and magnetic forces do not exist independently of the state of motion of the system of co-ordinates.

Furthermore it is clear that the asymmetry mentioned in the introduction as arising when we consider the currents produced by the relative motion of a magnet and a conductor, now disappears. Moreover, questions as to the "seat" of electrodynamic electromotive forces (unipolar machines) now have no point.

§ 7. Theory of Doppler's Principle and of Aberration

In the system K, very far from the origin of co-ordinates, let there be a source of electrodynamic waves, which in a part of space containing the origin of co-ordinates may be represented to a sufficient degree of approximation by the equations

$$X = X_0 \sin\Phi, \quad L = L_0 \sin\Phi,$$
$$Y = Y_0 \sin\Phi, \quad M = M_0 \sin\Phi,$$
$$Z = Z_0 \sin\Phi, \quad N = N_0 \sin\Phi,$$

where

$$\Phi = \omega\left\{t - \frac{1}{c}(lx + my + nz)\right\}.$$

Here (X_0, Y_0, Z_0) and (L_0, M_0, N_0) are the vectors defining the amplitude of the wave-train, and l, m, n the direction-cosines of the wave-normals. We wish to know the constitution of these waves, when they are examined by an observer at rest in the moving system k.

Applying the equations of transformation found in § 6 for electric and magnetic forces, and those found in § 3 for the co-ordinates and the time, we obtain directly

$$X' = X_0 \sin\Phi', \qquad L' = L_0 \sin\Phi',$$
$$Y' = \beta(Y_0 - vN_0/c)\sin\Phi', \quad M' = \beta(M_0 + vZ_0/c)\sin\Phi',$$
$$Z' = \beta(Z_0 + vM_0/c)\sin\Phi', \quad N' = \beta(N_0 - vY_0/c)\sin\Phi',$$
$$\Phi' = \omega'\left\{\tau - \frac{1}{c}(l'\xi + m'\eta + n'\zeta)\right\}$$

where

$$\omega' = \omega\beta(1 - lv/c),$$

$$l' = \frac{l - v/c}{1 - lv/c},$$

$$m' = \frac{m}{\beta(1 - lv/c)},$$

$$n' = \frac{n}{\beta(1 - lv/c)}.$$

From the equation for ω' it follows that if an observer is moving with velocity v relatively to an infinitely distant source of light of frequency ν, in such a way that the connecting line " source—observer " makes the angle ϕ with the velocity of the observer referred to a system of co-ordinates which is at rest relatively to the source of light, the frequency ν' of the light perceived by the observer is given by the equation

$$\nu' = \nu\frac{1 - \cos\phi \, . \, v/c}{\sqrt{(1 - v^2/c^2)}}.$$

This is Doppler's principle for any velocities whatever. When $\phi = 0$ the equation assumes the perspicuous form

$$\nu' = \nu\sqrt{\frac{1 - v/c}{1 + v/c}}.$$

We see that, in contrast with the customary view, when $v = - c, \nu' = \infty$.

If we call the angle between the wave-normal (direction of the ray) in the moving system and the connecting line " source—observer " ϕ', the equation for l' assumes the form

$$\cos\phi' = \frac{\cos\phi - v/c}{1 - \cos\phi \, . \, v/c}.$$

This equation expresses the law of aberration in its most general form. If $\phi = \frac{1}{2}\pi$, the equation becomes simply

$$\cos\phi' = - v/c.$$

We still have to find the amplitude of the waves, as it appears in the moving system. If we call the amplitude of the electric or magnetic force A or A′ respectively, accordingly as it is measured in the stationary system or in the moving system, we obtain

$$A'^2 = A^2 \frac{(1 - \cos\phi \,.\, v/c)^2}{1 - v^2/c^2}$$

which equation, if $\phi = 0$, simplifies into

$$A'^2 = A^2 \frac{1 - v/c}{1 + v/c}.$$

It follows from these results that to an observer approaching a source of light with the velocity c, this source of light must appear of infinite intensity.

§ 8. Transformation of the Energy of Light Rays. Theory of the Pressure of Radiation Exerted on Perfect Reflectors

Since $A^2/8\pi$ equals the energy of light per unit of volume, we have to regard $A'^2/8\pi$, by the principle of relativity, as the energy of light in the moving system. Thus A'^2/A^2 would be the ratio of the " measured in motion " to the " measured at rest " energy of a given light complex, if the volume of a light complex were the same, whether measured in K or in k. But this is not the case. If l, m, n are the direction-cosines of the wave-normals of the light in the stationary system, no energy passes through the surface elements of a spherical surface moving with the velocity of light:—

$$(x - lct)^2 + (y - mct)^2 + (z - nct)^2 = R^2.$$

We may therefore say that this surface permanently encloses the same light complex. We inquire as to the quantity of energy enclosed by this surface, viewed in system k, that is, as to the energy of the light complex relatively to the system k.

The spherical surface—viewed in the moving system—is an ellipsoidal surface, the equation for which, at the time $\tau = 0$, is

$$(\beta\xi - l\beta\xi v/c)^2 + (\eta - m\beta\xi v/c)^2 + (\zeta - n\beta\xi v/c)^2 = R^2.$$

If S is the volume of the sphere, and S′ that of this ellipsoid, then by a simple calculation

$$\frac{S'}{S} = \frac{\sqrt{1 - v^2/c^2}}{1 - \cos\phi \,.\, v/c}.$$

Thus, if we call the light energy enclosed by this surface E when it is measured in the stationary system, and E′ when measured in the moving system, we obtain

$$\frac{E'}{E} = \frac{A'^2S'}{A^2S} = \frac{1 - \cos\phi \,.\, v/c}{\sqrt{(1 - v^2/c^2)}},$$

and this formula, when $\phi = 0$, simplifies into

$$\frac{E'}{E} = \sqrt{\frac{1 - v/c}{1 + v/c}}.$$

It is remarkable that the energy and the frequency of a light complex vary with the state of motion of the observer in accordance with the same law.

Now let the co-ordinate plane $\xi = 0$ be a perfectly reflecting surface, at which the plane waves considered in § 7 are reflected. We seek for the pressure of light exerted on the reflecting surface, and for the direction, frequency, and intensity of the light after reflexion.

Let the incidental light be defined by the quantities A, $\cos\phi$, ν (referred to system K). Viewed from k the corresponding quantities are

$$A' = A\frac{1 - \cos\phi \,.\, v/c}{\sqrt{(1 - v^2/c^2)}},$$

$$\cos\phi' = \frac{\cos\phi - v/c}{1 - \cos\phi \,.\, v/c},$$

$$\nu' = \nu\frac{1 - \cos\phi \,.\, v/c}{\sqrt{(1 - v^2/c^2)}}.$$

For the reflected light, referring the process to system k, we obtain

$$A'' = A'$$
$$\cos\phi'' = -\cos\phi'$$
$$\nu'' = \nu'$$

Finally, by transforming back to the stationary system K, we obtain for the reflected light

$$A''' = A''\frac{1 + \cos\phi''\,.\,v/c}{\sqrt{(1 - v^2/c^2)}} = A\frac{1 - 2\cos\phi\,.\,v/c + v^2/c^2}{1 - v^2/c^2},$$

$$\cos\phi''' = \frac{\cos\phi'' + v/c}{1 + \cos\phi''\,.\,v/c} = -\frac{(1 + v^2/c^2)\cos\phi - 2v/c}{1 - 2\cos\phi\,.\,v/c + v^2/c^2}$$

$$\nu''' = \nu''\frac{1 + \cos\phi'' v/c}{\sqrt{(1 - v^2/c^2)}} = \nu\frac{1 - 2\cos\phi\,.\,v/c + v^2/c^2}{1 - v^2/c^2}.$$

The energy (measured in the stationary system) which is incident upon unit area of the mirror in unit time is evidently $A^2(c\cos\phi - v)/8\pi$. The energy leaving the unit of surface of the mirror in the unit of time is $A'''^2(-c\cos\phi''' + v)/8\pi$. The difference of these two expressions is, by the principle of energy, the work done by the pressure of light in the unit of time. If we set down this work as equal to the product Pv, where P is the pressure of light, we obtain

$$P = 2\,.\,\frac{A^2}{8\pi}\,\frac{(\cos\phi - v/c)^2}{1 - v^2/c^2}.$$

In agreement with experiment and with other theories, we obtain to a first approximation

$$P = 2\,.\,\frac{A^2}{8\pi}\cos^2\phi.$$

All problems in the optics of moving bodies can be solved by the method here employed. What is essential is, that the electric and magnetic force of the light which is influenced by a moving body, be transformed into a system of co-ordinates at rest relatively to the body. By this means all problems in the optics of moving bodies will be reduced to a series of problems in the optics of stationary bodies.

§ 9. Transformation of the Maxwell-Hertz Equations when Convection-Currents are Taken into Account

We start from the equations

$$\frac{1}{c}\left\{\frac{\partial X}{\partial t} + u_x\rho\right\} = \frac{\partial N}{\partial y} - \frac{\partial M}{\partial z}, \quad \frac{1}{c}\frac{\partial L}{\partial t} = \frac{\partial Y}{\partial z} - \frac{\partial Z}{\partial y},$$

$$\frac{1}{c}\left\{\frac{\partial Y}{\partial t} + u_y\rho\right\} = \frac{\partial L}{\partial z} - \frac{\partial N}{\partial x}, \quad \frac{1}{c}\frac{\partial M}{\partial t} = \frac{\partial Z}{\partial x} - \frac{\partial X}{\partial z},$$

$$\frac{1}{c}\left\{\frac{\partial Z}{\partial t} + u_z\rho\right\} = \frac{\partial M}{\partial x} - \frac{\partial L}{\partial y}, \quad \frac{1}{c}\frac{\partial N}{\partial t} = \frac{\partial X}{\partial y} - \frac{\partial Y}{\partial x},$$

where

$$\rho = \frac{\partial X}{\partial x} + \frac{\partial Y}{\partial y} + \frac{\partial Z}{\partial z}$$

denotes 4π times the density of electricity, and (u_x, u_y, u_z) the velocity-vector of the charge. If we imagine the electric charges to be invariably coupled to small rigid bodies (ions, electrons), these equations are the electromagnetic basis of the Lorentzian electrodynamics and optics of moving bodies.

Let these equations be valid in the system K, and transform them, with the assistance of the equations of transformation given in §§ 3 and 6, to the system k. We then obtain the equations

$$\frac{1}{c}\left\{\frac{\partial X'}{\partial \tau} + u_\xi \rho'\right\} = \frac{\partial N'}{\partial \eta} - \frac{\partial M'}{\partial \zeta}, \quad \frac{1}{c}\frac{\partial L'}{\partial \tau} = \frac{\partial Y'}{\partial \zeta} - \frac{\partial Z'}{\partial \eta},$$

$$\frac{1}{c}\left\{\frac{\partial Y'}{\partial \tau} + u_\eta \rho'\right\} = \frac{\partial L'}{\partial \zeta} - \frac{\partial N'}{\partial \xi}, \quad \frac{1}{c}\frac{\partial M'}{\partial \tau} = \frac{\partial Z'}{\partial \xi} - \frac{\partial X'}{\partial \zeta},$$

$$\frac{1}{c}\left\{\frac{\partial Z'}{\partial \tau} + u_\zeta \rho'\right\} = \frac{\partial M'}{\partial \xi} - \frac{\partial L'}{\partial \eta}, \quad \frac{1}{c}\frac{\partial N'}{\partial \tau} = \frac{\partial X'}{\partial \eta} - \frac{\partial Y'}{\partial \xi},$$

where

$$u_\xi = \frac{u_x - v}{1 - u_x v/c^2}$$

$$u_\eta = \frac{u_y}{\beta(1 - u_x v/c^2)}$$

$$u_\zeta = \frac{u_z}{\beta(1 - u_x v/c^2)},$$

and

$$\rho' = \frac{\partial X'}{\partial \xi} + \frac{\partial Y'}{\partial \eta} + \frac{\partial Z'}{\partial \zeta}$$

$$= \beta(1 - u_x v/c^2)\rho.$$

Since—as follows from the theorem of addition of velocities (§ 5)—the vector (u_ξ, u_η, u_ζ) is nothing else than the velocity of the electric charge, measured in the system k, we have the proof that, on the basis of our kinematical principles, the electrodynamic foundation of Lorentz's theory of the electrodynamics of moving bodies is in agreement with the principle of relativity.

In addition I may briefly remark that the following important law may easily be deduced from the developed equations: If an electrically charged body is in motion anywhere in space without altering its charge when regarded from a system of co-ordinates moving with the body, its charge also remains—when regarded from the "stationary" system K—constant.

§ 10. Dynamics of the Slowly Accelerated Electron

Let there be in motion in an electromagnetic field an electrically charged particle (in the sequel called an "electron"), for the law of motion of which we assume as follows:—

If the electron is at rest at a given epoch, the motion of the electron ensues in the next instant of time according to the equations

$$m\frac{d^2x}{dt^2} = \epsilon\mathrm{X}$$

$$m\frac{d^2y}{dt^2} = \epsilon\mathrm{Y}$$

$$m\frac{d^2z}{dt^2} = \epsilon\mathrm{Z}$$

where x, y, z denote the co-ordinates of the electron, and m the mass of the electron, as long as its motion is slow.

Now, secondly, let the velocity of the electron at a given epoch be v. We seek the law of motion of the electron in the immediately ensuing instants of time.

Without affecting the general character of our considerations, we may and will assume that the electron, at the moment when we give it our attention, is at the origin of the co-ordinates, and moves with the velocity v along the axis of X of the system K. It is then clear that at the given moment ($t = 0$) the electron is at rest relatively to a system of co-ordinates which is in parallel motion with velocity v along the axis of X.

From the above assumption, in combination with the principle of relativity, it is clear that in the immediately ensuing time (for small values of t) the electron, viewed from the system k, moves in accordance with the equations

$$m\frac{d^2\xi}{d\tau^2} = \epsilon X',$$

$$m\frac{d^2\eta}{d\tau^2} = \epsilon Y',$$

$$m\frac{d^2\zeta}{d\tau^2} = \epsilon Z',$$

in which the symbols ξ, η, ζ, τ, X′, Y′, Z′ refer to the system k. If, further, we decide that when $t = x = y = z = 0$ then $\tau = \xi = \eta = \zeta = 0$, the transformation equations of §§ 3 and 6 hold good, so that we have

$$\xi = \beta(x - vt),\ \eta = y,\ \zeta = z,\ \tau = \beta(t - vx/c^2)$$
$$X' = X,\ Y' = \beta(Y - vN/c),\ Z' = \beta(Z + vM/c).$$

With the help of these equations we transform the above equations of motion from system k to system K, and obtain

$$\left.\begin{aligned} \frac{d^2x}{dt^2} &= \frac{\epsilon}{m}\frac{1}{\beta^3}X \\ \frac{d^2y}{dt^2} &= \frac{\epsilon}{m}\frac{1}{\beta}\left(Y - \frac{v}{c}N\right) \\ \frac{d^2z}{dt^2} &= \frac{\epsilon}{m}\frac{1}{\beta}\left(Z + \frac{v}{c}M\right) \end{aligned}\right\} \quad . \quad . \quad . \quad (A)$$

Taking the ordinary point of view we now inquire as to the "longitudinal" and the "transverse" mass of the moving electron. We write the equations (A) in the form

$$m\beta^3\frac{d^2x}{dt^2} = \epsilon X = \epsilon X',$$

$$m\beta^2\frac{d^2y}{dt^2} = \epsilon\beta\left(Y - \frac{v}{c}N\right) = \epsilon Y',$$

$$m\beta^2\frac{d^2z}{dt^2} = \epsilon\beta\left(Z + \frac{v}{c}M\right) = \epsilon Z',$$

and remark firstly that $\epsilon X'$, $\epsilon Y'$, $\epsilon Z'$ are the components of the ponderomotive force acting upon the electron, and are so indeed as viewed in a system moving at the moment with the electron, with the same velocity as the electron. (This force might be measured, for example, by a spring balance at rest in the last-mentioned system.) Now if we call this force simply "the force acting upon the electron,"* and maintain the equation—mass × acceleration = force—and if we also decide that the accelerations are to be measured in the stationary system K, we derive from the above equations

$$\text{Longitudinal mass} = \frac{m}{(\sqrt{1 - v^2/c^2})^3}.$$

$$\text{Transverse mass} = \frac{m}{1 - v^2/c^2}.$$

With a different definition of force and acceleration we should naturally obtain other values for the masses. This shows us that in comparing different theories of the motion of the electron we must proceed very cautiously.

We remark that these results as to the mass are also valid for ponderable material points, because a ponderable material point can be made into an electron (in our sense of the word) by the addition of an electric charge, *no matter how small.*

We will now determine the kinetic energy of the electron. If an electron moves from rest at the origin of co-ordinates of the system K along the axis of X under the action of an electrostatic force X, it is clear that the energy withdrawn from the electrostatic field has the value $\int \epsilon \mathrm{X} dx$. As the electron is to be slowly accelerated, and consequently may not give off any energy in the form of radiation, the energy withdrawn from the electrostatic field must be put down as equal to the energy of motion W of the electron. Bearing in mind that during the whole process of motion which we are considering, the first of the equations (A) applies, we therefore obtain

$$\mathrm{W} = \int \epsilon \mathrm{X} dx = m\int_0^v \beta^3 v dv$$
$$= mc^2\left\{\frac{1}{\sqrt{1 - v^2/c^2}} - 1\right\}.$$

Thus, when $v = c$, W becomes infinite. Velocities

* The definition of force here given is not advantageous, as was first shown by M. Planck. It is more to the point to define force in such a way that the laws of momentum and energy assume the simplest form.

greater than that of light have—as in our previous results—no possibility of existence.

This expression for the kinetic energy must also, by virtue of the argument stated above, apply to ponderable masses as well.

We will now enumerate the properties of the motion of the electron which result from the system of equations (A), and are accessible to experiment.

1. From the second equation of the system (A) it follows that an electric force Y and a magnetic force N have an equally strong deflective action on an electron moving with the velocity v, when $Y = Nv/c$. Thus we see that it is possible by our theory to determine the velocity of the electron from the ratio of the magnetic power of deflexion A_m to the electric power of deflexion A_e, for any velocity, by applying the law

$$\frac{A_m}{A_e} = \frac{v}{c}.$$

This relationship may be tested experimentally, since the velocity of the electron can be directly measured, e.g. by means of rapidly oscillating electric and magnetic fields.

2. From the deduction for the kinetic energy of the electron it follows that between the potential difference, P, traversed and the acquired velocity v of the electron there must be the relationship

$$P = \int X dx = \frac{m}{\epsilon}c^2 \left\{ \frac{1}{\sqrt{1 - v^2/c^2}} - 1 \right\}$$

3. We calculate the radius of curvature of the path of the electron when a magnetic force N is present (as the only deflective force), acting perpendicularly to the velocity of the electron. From the second of the equations (A) we obtain

$$-\frac{d^2y}{dt^2} = \frac{v^2}{R} = \frac{\epsilon}{m}\frac{v}{c}N\sqrt{1 - \frac{v^2}{c^2}}$$

or

$$R = \frac{mc^2}{\epsilon} \cdot \frac{v/c}{\sqrt{(1 - v^2/c^2)}} \cdot \frac{1}{N}.$$

These three relationships are a complete expression for

the laws according to which, by the theory here advanced, the electron must move.

In conclusion I wish to say that in working at the problem here dealt with I have had the loyal assistance of my friend and colleague M. Besso, and that I am indebted to him for several valuable suggestions.

DOES THE INERTIA OF A BODY DEPEND UPON ITS ENERGY-CONTENT?

BY

A. EINSTEIN

Translated from "Ist die Trägheit eines Körpers von seinem Energiegehalt abhängig?" Annalen der Physik, 17, 1905.

DOES THE INERTIA OF A BODY DEPEND UPON ITS ENERGY-CONTENT?

By A. EINSTEIN

THE results of the previous investigation lead to a very interesting conclusion, which is here to be deduced.

I based that investigation on the Maxwell-Hertz equations for empty space, together with the Maxwellian expression for the electromagnetic energy of space, and in addition the principle that:—

The laws by which the states of physical systems alter are independent of the alternative, to which of two systems of co-ordinates, in uniform motion of parallel translation relatively to each other, these alterations of state are referred (principle of relativity).

With these principles * as my basis I deduced *inter alia* the following result (§ 8):—

Let a system of plane waves of light, referred to the system of co-ordinates (x, y, z), possess the energy l; let the direction of the ray (the wave-normal) make an angle ϕ with the axis of x of the system. If we introduce a new system of co-ordinates (ξ, η, ζ) moving in uniform parallel translation with respect to the system (x, y, z), and having its origin of co-ordinates in motion along the axis of x with the velocity v, then this quantity of light—measured in the system (ξ, η, ζ) —possesses the energy

$$l^* = l \frac{1 - \frac{v}{c} \cos \phi}{\sqrt{1 - v^2/c^2}}$$

* The principle of the constancy of the velocity of light is of course contained in Maxwell's equations.

where c denotes the velocity of light. We shall make use of this result in what follows.

Let there be a stationary body in the system (x, y, z), and let its energy—referred to the system (x, y, z)—be E_0. Let the energy of the body relative to the system (ξ, η, ζ), moving as above with the velocity v, be H_0.

Let this body send out, in a direction making an angle ϕ with the axis of x, plane waves of light, of energy $\frac{1}{2}L$ measured relatively to (x, y, z), and simultaneously an equal quantity of light in the opposite direction. Meanwhile the body remains at rest with respect to the system (x, y, z). The principle of energy must apply to this process, and in fact (by the principle of relativity) with respect to both systems of co-ordinates. If we call the energy of the body after the emission of light E_1 or H_1 respectively, measured relatively to the system (x, y, z) or (ξ, η, ζ) respectively, then by employing the relation given above we obtain

$$E_0 = E_1 + \tfrac{1}{2}L + \tfrac{1}{2}L,$$

$$H_0 = H_1 + \tfrac{1}{2}L\frac{1 - \frac{v}{c}\cos\phi}{\sqrt{1 - v^2/c^2}} + \tfrac{1}{2}L\frac{1 + \frac{v}{c}\cos\phi}{\sqrt{1 - v^2/c^2}}$$

$$= H_1 + \frac{L}{\sqrt{1 - v^2/c^2}}.$$

By subtraction we obtain from these equations

$$H_0 - E_0 - (H_1 - E_1) = L\left\{\frac{1}{\sqrt{1 - v^2/c^2}} - 1\right\}.$$

The two differences of the form $H - E$ occurring in this expression have simple physical significations. H and E are energy values of the same body referred to two systems of co-ordinates which are in motion relatively to each other, the body being at rest in one of the two systems (system (x, y, z)). Thus it is clear that the difference $H - E$ can differ from the kinetic energy K of the body, with respect to the other system (ξ, η, ζ), only by an additive constant C, which depends on the choice of the arbitrary additive constants of the energies H and E. Thus we may place

$$H_0 - E_0 = K_0 + C,$$
$$H_1 - E_1 = K_1 + C,$$

since C does not change during the emission of light. So we have

$$K_0 - K_1 = L\left\{\frac{1}{\sqrt{1 - v^2/c^2}} - 1\right\}.$$

The kinetic energy of the body with respect to (ξ, η, ζ) diminishes as a result of the emission of light, and the amount of diminution is independent of the properties of the body. Moreover, the difference $K_0 - K_1$, like the kinetic energy of the electron (§ 10), depends on the velocity.

Neglecting magnitudes of fourth and higher orders we may place

$$K_0 - K_1 = \frac{1}{2}\frac{L}{c^2}v^2.$$

From this equation it directly follows that :—

If a body gives off the energy L in the form of radiation, its mass diminishes by L/c^2. The fact that the energy withdrawn from the body becomes energy of radiation evidently makes no difference, so that we are led to the more general conclusion that

The mass of a body is a measure of its energy-content; if the energy changes by L, the mass changes in the same sense by $L/9 \times 10^{20}$, the energy being measured in ergs, and the mass in grammes.

It is not impossible that with bodies whose energy-content is variable to a high degree (e.g. with radium salts) the theory may be successfully put to the test.

If the theory corresponds to the facts, radiation conveys inertia between the emitting and absorbing bodies.

SPACE AND TIME

BY

H. MINKOWSKI

A Translation of an Address delivered at the 80*th Assembly of German Natural Scientists and Physicians, at Cologne,* 21 *September*, 1908.

SPACE AND TIME

By H. MINKOWSKI

THE views of space and time which I wish to lay before you have sprung from the soil of experimental physics, and therein lies their strength. They are radical. Henceforth space by itself, and time by itself, are doomed to fade away into mere shadows, and only a kind of union of the two will preserve an independent reality.

I

First of all I should like to show how it might be possible, setting out from the accepted mechanics of the present day, along a purely mathematical line of thought, to arrive at changed ideas of space and time. The equations of Newton's mechanics exhibit a two-fold invariance. Their form remains unaltered, firstly, if we subject the underlying system of spatial co-ordinates to any arbitrary *change of position;* secondly, if we change its state of motion, namely, by imparting to it any *uniform translatory motion;* furthermore, the zero point of time is given no part to play. We are accustomed to look upon the axioms of geometry as finished with, when we feel ripe for the axioms of mechanics, and for that reason the two invariances are probably rarely mentioned in the same breath. Each of them by itself signifies, for the differential equations of mechanics, a certain group of transformations. The existence of the first group is looked upon as a fundamental characteristic of space. The second group is preferably treated with disdain, so that we with untroubled minds may overcome the difficulty of never being able to decide, from physical phenomena, whether space, which is supposed to be stationary, may not be after all in a state of uniform translation. Thus the two groups, side by side, lead their lives entirely apart. Their utterly heterogeneous character may have discouraged any attempt to compound them. But it is precisely when they are compounded that the complete group, as a whole, gives us to think.

We will try to visualize the state of things by the graphic method. Let x, y, z be rectangular co-ordinates for space, and let t denote time. The objects of our perception invariably include places and times in combination. Nobody has ever noticed a place except at a time, or a time except at a place. But I still respect the dogma that both space and time have independent significance. A point of space at a point of time, that is, a system of values x, y, z, t, I will call a *world-point*. The multiplicity of all thinkable x, y, z, t systems of values we will christen the *world*. With this most valiant piece of chalk I might project upon the blackboard four world-axes. Since merely one chalky axis, as it is, consists of molecules all a-thrill, and moreover is taking part in the earth's travels in the universe, it already affords us ample scope for abstraction; the somewhat greater abstraction associated with the number four is for the mathematician no infliction. Not to leave a yawning void anywhere, we will imagine that everywhere and everywhen there is something perceptible. To avoid saying "matter" or "electricity" I will use for this something the word "substance." We fix our attention on the substantial point which is at the world-point x, y, z, t, and imagine that we are able to recognize this substantial point at any other time. Let the variations dx, dy, dz of the space co-ordinates of this substantial point correspond to a time element dt. Then we obtain, as an image, so to speak, of the everlasting career of the substantial point, a curve in the world, a *world-line*, the points of which can be referred unequivocally to the parameter t from $-\infty$ to $+\infty$. The whole universe is seen to resolve itself into similar world-lines, and I would fain anticipate myself by saying that in my opinion physical laws might find their most perfect expression as reciprocal relations between these world-lines.

The concepts, space and time, cause the x, y, z-manifold $t = 0$ and its two sides $t > 0$ and $t < 0$ to fall asunder. If, for simplicity, we retain the same zero point of space and time, the first-mentioned group signifies in mechanics that we may subject the axes of x, y, z at $t = 0$ to any rotation we choose about the origin, corresponding to the homogeneous linear transformations of the expression

$$x^2 + y^2 + z^2.$$

But the second group means that we may—also without changing the expression of the laws of mechanics—replace x, y, z, t by $x - at, y - \beta t, z - \gamma t, t$ with any constant values of a, β, γ. Hence we may give to the time axis whatever direction we choose towards the upper half of the world, $t > 0$. Now what has the requirement of orthogonality in space to do with this perfect freedom of the time axis in an upward direction?

To establish the connexion, let us take a positive parameter c, and consider the graphical representation of

$$c^2t^2 - x^2 - y^2 - z^2 = 1.$$

It consists of two surfaces separated by $t = 0$, on the analogy of a hyperboloid of two sheets. We consider the sheet in the region $t > 0$, and now take those homogeneous linear transformations of x, y, z, t into four new variables x', y', z', t', for which the expression for this sheet in the new variables is of the same form. It is evident that the rotations of space about the origin pertain to these transformations. Thus we gain full comprehension of the rest of the transformations simply by taking into consideration one among them, such that y and z remain unchanged. We draw (Fig. 1) the section of this sheet by the plane of the axes of x and t—the upper branch of the hyperbola $c^2t^2 - x^2 = 1$, with its asymptotes. From the origin O we draw any radius vector OA′ of this branch of the hyperbola; draw the tangent to the hyperbola at A′ to cut the asymptote on the right at B′; complete the parallelogram OA′B′C′; and finally, for subsequent use, produce B′C′ to cut the axis of x at D′. Now if we take OC′ and OA′ as axes of oblique co-ordinates x', t', with the measures OC′ = 1, OA′ = $1/c$, then that branch of the hyperbola again acquires the expression $c^2t'^2 - x'^2 = 1$, $t' > 0$, and the transition from x, y, z, t to x', y', z', t' is one of the transformations in question. With these transformations we now associate the arbitrary displacements of the zero point of space and time, and thereby constitute a group of transformations, which is also, evidently, dependent on the parameter c. This group I denote by G_c.

If we now allow c to increase to infinity, and $1/c$ therefore to converge towards zero, we see from the figure that the

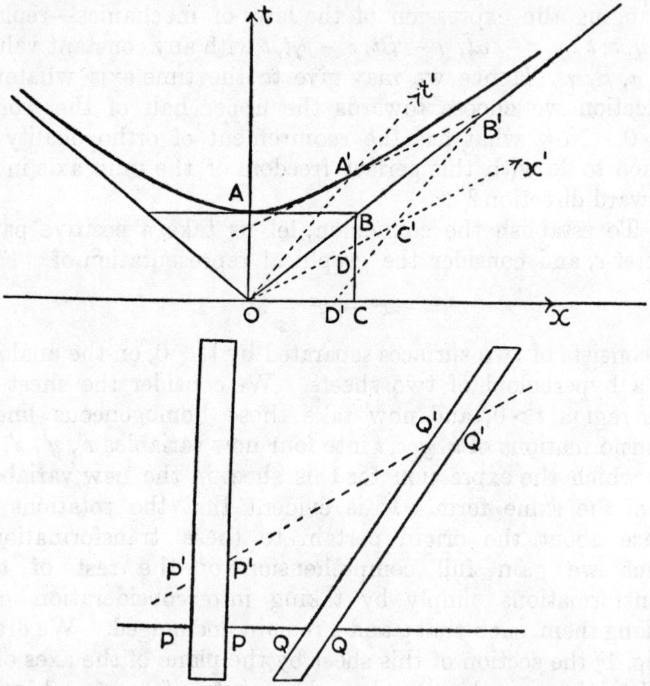

FIG. 1.

branch of the hyperbola bends more and more towards the axis of x, the angle of the asymptotes becomes more and more obtuse, and that in the limit this special transformation changes into one in which the axis of t' may have any upward direction whatever, while x' approaches more and more exactly to x. In view of this it is clear that group G_c in the limit when $c = \infty$, that is the group G_∞, becomes no other than that complete group which is appropriate to Newtonian mechanics. This being so, and since G_c is mathematically more intelligible than G_∞, it looks as though the thought might have struck some mathematician, fancy-free, that after all, as a matter of fact, natural phenomena do not possess an invariance with the group G_∞, but rather with a group G_c, c being finite and determinate, but in ordinary units of measure, *extremely great.* Such a premonition would have been an extraordinary triumph for pure mathematics. Well, mathematics, though it now can display only staircase-wit, has the satisfaction of being wise after the event, and is able, thanks to its happy antecedents, with its senses sharpened by an unhampered outlook to far horizons, to grasp forthwith the far-reaching consequences of such a metamorphosis of our concept of nature.

I will state at once what is the value of c with which we shall finally be dealing. It is the velocity of the propagation of light in empty space. To avoid speaking either of space or of emptiness, we may define this magnitude in another way, as the ratio of the electromagnetic to the electrostatic unit of electricity.

The existence of the invariance of natural laws for the relevant group G_c would have to be taken, then, in this way:—

From the totality of natural phenomena it is possible, by successively enhanced approximations, to derive more and more exactly a system of reference x, y, z, t, space and time, by means of which these phenomena then present themselves in agreement with definite laws. But when this is done, this system of reference is by no means unequivocally determined by the phenomena. *It is still possible to make any change in the system of reference that is in conformity with the transformations of the group G_c, and leave the expression of the laws of nature unaltered.*

For example, in correspondence with the figure described above, we may also designate time t', but then must of necessity, in connexion therewith, define space by the manifold of the three parameters x', y, z, in which case physical laws would be expressed in exactly the same way by means of x', y, z, t' as by means of x, y, z, t. We should then have in the world no longer *space*, but an infinite number of spaces, analogously as there are in three-dimensional space an infinite number of planes. Three-dimensional geometry becomes a chapter in four-dimensional physics. Now you know why I said at the outset that space and time are to fade away into shadows, and only a world in itself will subsist.

II

The question now is, what are the circumstances which force this changed conception of space and time upon us? Does it actually never contradict experience? And finally, is it advantageous for describing phenomena?

Before going into these questions, I must make an important remark. If we have in any way individualized space and time, we have, as a world-line corresponding to a stationary substantial point, a straight line parallel to the axis of t; corresponding to a substantial point in uniform motion, a straight line at an angle to the axis of t; to a substantial point in varying motion, a world-line in some form of a curve. If at any world-point x, y, z, t we take the world-line passing through that point, and find it parallel to any radius vector OA′ of the above-mentioned hyperboloidal sheet, we can introduce OA′ as a new axis of time, and with the new concepts of space and time thus given, the substance at the world-point concerned appears as at rest. We will now introduce this fundamental axiom :—

The substance at any world-point may always, with the appropriate determination of space and time, be looked upon as at rest.

The axiom signifies that at any world-point the expression

$$c^2dt^2 - dx^2 - dy^2 - dz^2$$

always has a positive value, or, what comes to the same thing, that any velocity v always proves less than c. Accordingly c would stand as the upper limit for all substantial velocities, and that is precisely what would reveal the deeper significance of the magnitude c. In this second form the first impression made by the axiom is not altogether pleasing. But we must bear in mind that a modified form of mechanics, in which the square root of this quadratic differential expression appears, will now make its way, so that cases with a velocity greater than that of light will henceforward play only some such part as that of figures with imaginary co-ordinates in geometry.

Now the impulse and true motive for assuming the group G_c came from the fact that the differential equation for the propagation of light in empty space possesses that group G_c.* On the other hand, the concept of rigid bodies has meaning only in mechanics satisfying the group G_∞. If we have a theory of optics with G_c, and if on the other hand there were rigid bodies, it is easy to see that one and the same direction of t would be distinguished by the two hyperboloidal sheets appropriate to G_c and G_∞, and this would have the further consequence, that we should be able, by employing suitable rigid optical instruments in the laboratory, to perceive some alteration in the phenomena when the orientation with respect to the direction of the earth's motion is changed. But all efforts directed towards this goal, in particular the famous interference experiment of Michelson, have had a negative result. To explain this failure, H. A. Lorentz set up an hypothesis, the success of which lies in this very invariance in optics for the group G_c. According to Lorentz any moving body must have undergone a contraction in the direction of its motion, and in fact with a velocity v, a contraction in the ratio

$$1 : \sqrt{1 - v^2/c^2}.$$

This hypothesis sounds extremely fantastical, for the contraction is not to be looked upon as a consequence of resistances in the ether, or anything of that kind, but simply as a gift from above,—as an accompanying circumstance of the circumstance of motion.

I will now show by our figure that the Lorentzian hypothesis is completely equivalent to the new conception of space and time, which, indeed, makes the hypothesis much more intelligible. If for simplicity we disregard y and z, and imagine a world of one spatial dimension, then a parallel band, upright like the axis of t, and another inclining to the axis of t (see Fig. 1)

* An application of this fact in its essentials has already been given by W. Voigt, Gottinger Nachrichten, 1887, p. 41.

represent, respectively, the career of a body at rest or in uniform motion, preserving in each case a constant spatial extent. If OA′ is parallel to the second band, we can introduce t' as the time, and x' as the space co-ordinate, and then the second body appears at rest, the first in uniform motion. We now assume that the first body, envisaged as at rest, has the length l, that is, the cross section PP of the first band on the axis of x is equal to $l\,.\,\mathrm{OC}$, where OC denotes the unit of measure on the axis of x; and on the other hand, that the second body, envisaged as at rest, has the same length l, which then means that the cross section Q′Q′ of the second band, measured parallel to the axis of x', is equal to $l\,.\,\mathrm{OC}'$. We now have in these two bodies images of two equal Lorentzian electrons, one at rest and one in uniform motion. But if we retain the original co-ordinates x, t, we must give as the extent of the second electron the cross section of its appropriate band parallel to the axis of x. Now since Q′Q′ $= l\,.\,\mathrm{OC}'$, it is evident that QQ $= l\,.\,\mathrm{OD}'$. If dx/dt for the second band is equal to v, an easy calculation gives

$$\mathrm{OD}' = \mathrm{OC}\sqrt{1 - v^2/c^2},$$

therefore also $\mathrm{PP} : \mathrm{QQ} = 1 : \sqrt{1 - v^2/c^2}$. But this is the meaning of Lorentz's hypothesis of the contraction of electrons in motion. If on the other hand we envisage the second electron as at rest, and therefore adopt the system of reference $x'\,t'$, the length of the first must be denoted by the cross section P′P′ of its band parallel to OC′, and we should find the first electron in comparison with the second to be contracted in exactly the same proportion; for in the figure

$$\mathrm{P'P'} : \mathrm{Q'Q'} = \mathrm{OD} : \mathrm{OC}' = \mathrm{OD}' : \mathrm{OC} = \mathrm{QQ} : \mathrm{PP}.$$

Lorentz called the t' combination of x and t the local time of the electron in uniform motion, and applied a physical construction of this concept, for the better understanding of the hypothesis of contraction. But the credit of first recognizing clearly that the time of the one electron is just as good as that of the other, that is to say, that t and t' are to be treated identically, belongs to A. Einstein.* Thus time, as a

* A. Einstein, Ann. d. Phys., 17, 1905, p. 891; Jahrb. d. Radioaktivität und Elektronik, 4, 1907, p. 411.

concept unequivocally determined by phenomena, was first deposed from its high seat. Neither Einstein nor Lorentz made any attack on the concept of space, perhaps because in the above-mentioned special transformation, where the plane of x', t' coincides with the plane of x, t, an interpretation is possible by saying that the x-axis of space maintains its position. One may expect to find a corresponding violation of the concept of space appraised as another act of audacity on the part of the higher mathematics. Nevertheless, this further step is indispensable for the true understanding of the group G_c, and when it has been taken, the word *relativity-postulate* for the requirement of an invariance with the group G_c seems to me very feeble. Since the postulate comes to mean that only the four-dimensional world in space and time is given by phenomena, but that the projection in space and in time may still be undertaken with a certain degree of freedom, I prefer to call it the *postulate of the absolute world* (or briefly, the world-postulate).

III

The world-postulate permits identical treatment of the four co-ordinates x, y, z, t. By this means, as I shall now show, the forms in which the laws of physics are displayed gain in intelligibility. In particular the idea of acceleration acquires a clear-cut character.

I will use a geometrical manner of expression, which suggests itself at once if we tacitly disregard z in the triplex x, y, z. I take any world-point O as the zero-point of space-time. The cone $c^2t^2 - x^2 - y^2 - z^2 = 0$ with apex O (Fig. 2) consists of two parts, one with values $t < 0$, the other with values $t > 0$. The former, the front cone of O, consists, let us say, of all the world-points which "send light to O," the latter, the back cone of O, of all the world-points which "receive light from O." The territory bounded by the front cone alone, we may call "before" O, that which is bounded by the back cone alone, "after" O. The hyperboloidal sheet already discussed

$$F = c^2t^2 - x^2 - y^2 - z^2 = 1, \quad t > 0$$

lies after O. The territory between the cones is filled by the one-sheeted hyperboloidal figures

$$-\mathrm{F} = x^2 + y^2 + z^2 - c^2t^2 = k^2$$

for all constant positive values of k. We are specially interested in the hyperbolas with O as centre, lying on the latter figures. The single branches of these hyperbolas may be called briefly the internal hyperbolas with centre O. One of these branches, regarded as a world-line, would represent a motion which, for $t = -\infty$ and $t = +\infty$, rises asymptotically to the velocity of light, c.

If we now, on the analogy of vectors in space, call a directed length in the manifold of x, y, z, t a vector, we have to distinguish between the time-like vectors with directions from O to the sheet $+\mathrm{F} = 1, t > 0$, and the space-like vectors

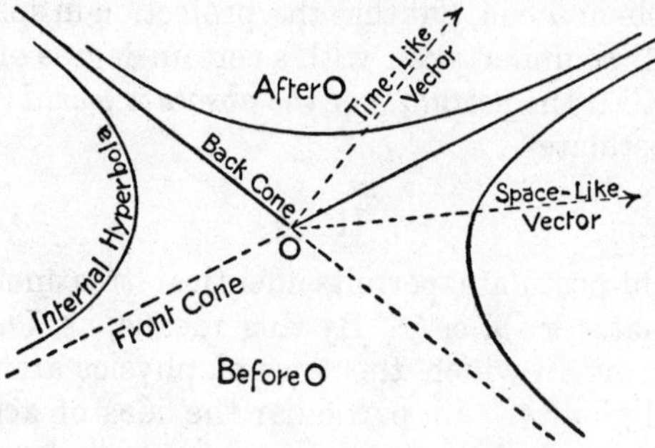

FIG. 2.

with directions from O to $-\mathrm{F} = 1$. The time axis may run parallel to any vector of the former kind. Any world-point between the front and back cones of O can be arranged by means of the system of reference so as to be simultaneous with O, but also just as well so as to be earlier than O or later than O. Any world-point within the front cone of O is necessarily always before O; any world-point within the back cone of O necessarily always after O. Corresponding to passing to the limit, $c = \infty$, there would be a complete flattening out of the wedge-shaped segment between the cones into the plane manifold $t = 0$. In the figures this segment is intentionally drawn with different widths.

We divide up any vector we choose, e.g. that from O to x, y, z, t, into the four components x, y, z, t. If the directions of two vectors are, respectively, that of a radius vector OR from O to one of the surfaces $\mp F = 1$, and that of a tangent RS at the point R of the same surface, the vectors are said to be normal to one another. Thus the condition that the vectors with components x, y, z, t and x_1, y_1, z_1, t_1 may be normal to each other is

$$c^2tt_1 - xx_1 - yy_1 - zz_1 = 0.$$

For the measurement of vectors in different directions the units of measure are to be fixed by assigning to a space-like vector from O to $- F = 1$ always the magnitude 1, and to a time-like vector from O to $+ F = 1, t > 0$ always the magnitude $1/c$.

If we imagine at a world-point P (x, y, z, t) the world-line of a substantial point running through that point, the magnitude corresponding to the time-like vector dx, dy, dz, dt laid off along the line is therefore

$$d\tau = \frac{1}{c}\sqrt{c^2dt^2 - dx^2 - dy^2 - dz^2}.$$

The integral $\int d\tau = \tau$ of this amount, taken along the world-line from any fixed starting-point P_0 to the variable end-point P, we call the proper time of the substantial point at P. On the world-line we regard x, y, z, t—the components of the vector OP—as functions of the proper time τ; denote their first differential coefficients with respect to τ by $\dot{x}, \dot{y}, \dot{z}, \dot{t}$; their second differential coefficients with respect to τ by $\ddot{x}, \ddot{y}, \ddot{z}, \ddot{t}$; and give names to the appropriate vectors, calling the derivative of the vector OP with respect to τ the velocity vector at P, and the derivative of this velocity vector with respect to τ the acceleration vector at P. Hence, since

$$c^2\dot{t}^2 - \dot{x}^2 - \dot{y}^2 - \dot{z}^2 = c^2,$$

we have

$$c^2\dot{t}\ddot{t} - \dot{x}\ddot{x} - \dot{y}\ddot{y} - \dot{z}\ddot{z} = 0,$$

i.e. the velocity vector is the time-like vector of unit magnitude in the direction of the world-line at P, and the acceleration vector at P is normal to the velocity vector at P, and is therefore in any case a space-like vector.

Now, as is readily seen, there is a definite hyperbola

which has three infinitely proximate points in common with the world-line at P, and whose asymptotes are generators of a "front cone" and a "back cone" (Fig. 3).

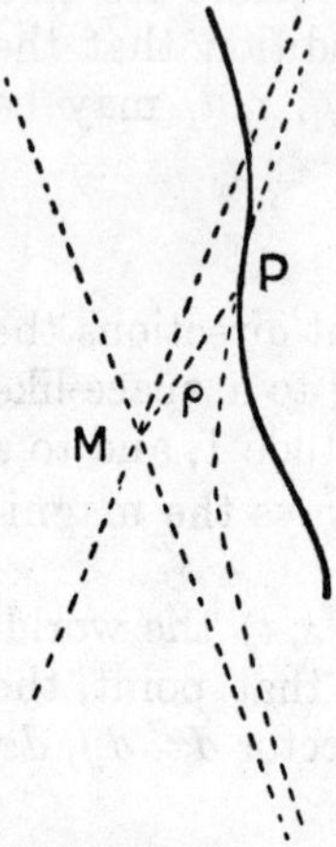

Fig. 3.

Let this hyperbola be called the hyperbola of curvature at P. If M is the centre of this hyperbola, we here have to do with an internal hyperbola with centre M. Let ρ be the magnitude of the vector MP; then we recognize the acceleration vector at P as the vector in the direction MP of magnitude c^2/ρ.

If $\ddot{x}, \ddot{y}, \ddot{z}, \ddot{t}$ are all zero, the hyperbola of curvature reduces to the straight line touching the world-line in P, and we must put $\rho = \infty$.

IV

To show that the assumption of group G_c for the laws of physics never leads to a contradiction, it is unavoidable to undertake a revision of the whole of physics on the basis of this assumption. This revision has to some extent already been successfully carried out for questions of thermodynamics and heat radiation,* for electromagnetic processes, and finally, with the retention of the concept of mass, for mechanics.†

For this last branch of physics it is of prime importance to raise the question—When a force with the components X, Y, Z parallel to the axes of space acts at a world-point P (x, y, z, t), where the velocity vector is $\dot{x}, \dot{y}, \dot{z}, \dot{t}$, what must we take this force to be when the system of reference is in any way changed? Now there exist certain approved statements as to the ponderomotive force in the electromagnetic field in the cases where the group G_c is undoubtedly admissible. These statements lead up to the simple rule:—When the system of reference is changed, the force in question transforms into a force in the new space co-ordinates in such a way that the appropriate vector with the components $\dot{t}X$,

* M. Planck, "Zur Dynamik bewegter Systeme," Berliner Berichte, 1907, p. 542; also in Ann. d. Phys., 26, 1908, p. 1.

† H. Minkowski, "Die Grundgleichungen für die elektromagnetischen Vorgänge in bewegten Körpern," Göttinger Nachrichten, 1908, p. 53.

iY, iZ, iT, where

$$T = \frac{1}{c^2}\left(\frac{\dot{x}}{t}X + \frac{\dot{y}}{t}Y + \frac{\dot{z}}{t}Z\right)$$

is the rate at which work is done by the force at the world-point divided by c, remains unchanged. This vector is always normal to the velocity vector at P. A force vector of this kind, corresponding to a force at P, is to be called a "motive force vector" at P.

I shall now describe the world-line of a substantial point with constant mechanical mass m, passing through P. Let the velocity vector at P, multiplied by m, be called the "momentum vector" at P, and the acceleration vector at P, multiplied by m, be called the "force vector" of the motion at P. With these definitions, the law of motion of a point of mass with given motive force vector runs thus:—* *The Force Vector of Motion is Equal to the Motive Force Vector.* This assertion comprises four equations for the components corresponding to the four axes, and since both vectors mentioned are *a priori* normal to the velocity vector, the fourth equation may be looked upon as a consequence of the other three. In accordance with the above signification of T, the fourth equation undoubtedly represents the law of energy. Therefore the component of the momentum vector along the axis of t, multiplied by c, is to be defined as the kinetic energy of the point mass. The expression for this is

$$mc^2\frac{dt}{d\tau} = mc^2/\sqrt{1 - v^2/c^2},$$

i.e., after removal of the additive constant mc^2, the expression $\frac{1}{2}mv^2$ of Newtonian mechanics down to magnitudes of the order $1/c^2$. It comes out very clearly in this way, how the energy depends on the system of reference. But as the axis of t may be laid in the direction of any time-like vector, the law of energy, framed for all possible systems of reference, already contains, on the other hand, the whole system of the equations of motion. At the limiting transition which we have discussed, to $c = \infty$, this fact retains its importance for

* H. Minkowski, loc. cit., p 107. Cf. also M. Planck, Verhandlungen der physikalischen Gesellschaft, 4, 1906, p. 136.

the axiomatic structure of Newtonian mechanics as well, and has already been appreciated in this sense by I. R. Schütz.*

We can determine the ratio of the units of length and time beforehand in such a way that the natural limit of velocity becomes $c = 1$. If we then introduce, further, $\sqrt{-1}\, t = s$ in place of t, the quadratic differential expression

$$d\tau^2 = -dx^2 - dy^2 - dz^2 - ds^2$$

thus becomes perfectly symmetrical in x, y, z, s; and this symmetry is communicated to any law which does not contradict the world-postulate. Thus the essence of this postulate may be clothed mathematically in a very pregnant manner in the mystic formula

$$3 \cdot 10^5\ km = \sqrt{-1}\ \text{secs.}$$

V

The advantages afforded by the world-postulate will perhaps be most strikingly exemplified by indicating the effects proceeding from a point charge in any kind of motion according to the Maxwell-Lorentz theory. Let us imagine the world-line of such a point electron with the charge e, and introduce upon it the proper time τ from any initial point. In order to find the field caused by the electron at any world-point P_1, we construct the front cone belonging to P_1 (Fig. 4). The cone evidently meets the world-line, since the directions of the line are everywhere those of time-like vectors, at the single point P. We draw the tangent to the world-line at P, and construct through P_1 the normal P_1Q to this tangent. Let the length of P_1Q be r. Then, by the definition of a front cone, the length of PQ must be r/c. Now the vector in the direction PQ of magnitude e/r repre-

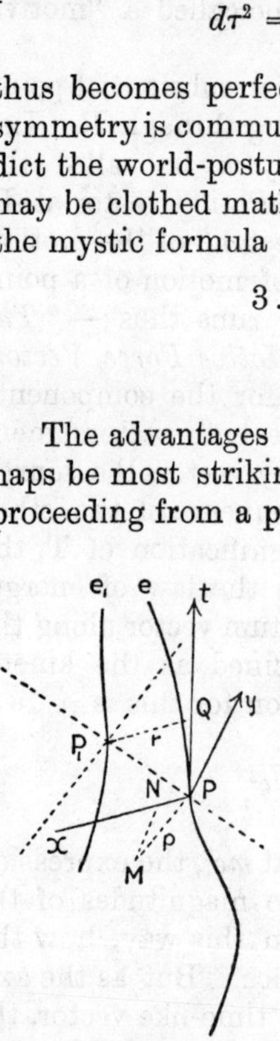

FIG. 4.

* I. R. Schütz, "Das Prinzip der absoluten Erhaltung der Energie," Göttinger Nachr., 1897, p. 110.

sents by its components along the axes of x, y, z, the vector potential multiplied by c, and by the component along the axis of t, the scalar potential of the field excited by e at the world-point P. Herein lie the elementary laws formulated by A. Liénard and E. Wiechert.*

Then in the description of the field produced by the electron we see that the separation of the field into electric and magnetic force is a relative one with regard to the underlying time axis; the most perspicuous way of describing the two forces together is on a certain analogy with the wrench in mechanics, though the analogy is not complete.

I will now describe the ponderomotive action of a moving point charge on another moving point charge. Let us imagine the world-line of a second point electron of the charge e_1, passing through the world-point P_1. We define P, Q, r as before, then construct (Fig. 4) the centre M of the hyperbola of curvature at P, and finally the normal MN from M to a straight line imagined through P parallel to QP_1. With P as starting-point we now determine a system of reference as follows:—The axis of t in the direction PQ, the axis of x in direction QP_1, the axis of y in direction MN, whereby finally the direction of the axis of z is also defined as normal to the axes of t, x, y. Let the acceleration vector at P be $\ddot{x}, \ddot{y}, \ddot{z}, \ddot{t}$, the velocity vector at P_1 be $\dot{x}_1, \dot{y}_1, \dot{z}_1, \dot{t}_1$. The motive force vector exerted at P_1 by the first moving electron e on the second moving electron e_1 now takes the form

$$-ee_1\left(\dot{t}_1 - \frac{\dot{x}_1}{c}\right)\mathfrak{K},$$

where the components $\mathfrak{K}_x$, $\mathfrak{K}_y$, $\mathfrak{K}_z$, $\mathfrak{K}_t$ of the vector $\mathfrak{K}$ satisfy the three relations

$$c\mathfrak{K}_t - \mathfrak{K}_x = \frac{1}{r^2}, \quad \mathfrak{K}_y = \frac{\ddot{y}}{c^2 r}, \quad \mathfrak{K}_z = 0,$$

and where, fourthly, this vector $\mathfrak{K}$ is normal to the velocity vector at P_1, and through this circumstance alone stands in dependence on the latter velocity vector.

* A. Liénard, "Champ électrique et magnétique produit par une charge concentrée en un point et animée d'un mouvement quelconque," L'Eclairage Electrique, 16, 1898, pp. 5, 53, 106; E. Wiechert, "Elektrodynamische Elementargesetze," Arch. Néerl. (2), 5, 1900, p. 549.

When we compare this statement with previous formulations* of the same elementary law of the ponderomotive action of moving point charges on one another, we are compelled to admit that it is only in four dimensions that the relations here taken under consideration reveal their inner being in full simplicity, and that on a three dimensional space forced upon us *a priori* they cast only a very complicated projection.

In mechanics as reformed in accordance with the world-postulate, the disturbing lack of harmony between Newtonian mechanics and modern electrodynamics disappears of its own accord. Before concluding I will just touch upon the attitude of Newton's law of attraction toward this postulate. I shall assume that when two points of mass m, m_1 describe their world-lines, a motive force vector is exerted by m on m_1, of exactly the same form as that just given in the case of electrons, except that $+ mm_1$ must now take the place of $- ee_1$. We now specially consider the case where the acceleration vector of m is constantly zero. Let us then introduce t in such a way that m is to be taken as at rest, and let only m_1 move under the motive force vector which proceeds from m. If we now modify this given vector in the first place by adding the factor $\dot{t}^{-1} = \sqrt{1 - v^2/c^2}$, which, to the order of $1/c^2$, is equal to 1, it will be seen † that for the positions x_1, y_1, z_1, of m_1 and their variations in time, we should arrive exactly at Kepler's laws again, except that the proper times τ_1 of m_1 would take the place of the times t_1. From this simple remark it may then be seen that the proposed law of attraction combined with the new mechanics is no less well adapted to explain astronomical observations than the Newtonian law of attraction combined with Newtonian mechanics.

The fundamental equations for electromagnetic processes in ponderable bodies also fit in completely with the world-postulate. As I shall show elsewhere, it is not even by any means necessary to abandon the derivation of these funda-

* K. Schwarzwald, Göttinger Nachr., 1903, p. 132; H. A. Lorentz, Enzykl. d. math. Wissensch., V, Art. 14, p. 199.

† H. Minkowski, loc. cit., p. 110.

mental equations from ideas of the electronic theory, as taught by Lorentz, in order to adapt them to the world-postulate.

The validity without exception of the world-postulate, I like to think, is the true nucleus of an electromagnetic image of the world, which, discovered by Lorentz, and further revealed by Einstein, now lies open in the full light of day. In the development of its mathematical consequences there will be ample suggestions for experimental verifications of the postulate, which will suffice to conciliate even those to whom the abandonment of old-established views is unsympathetic or painful, by the idea of a pre-established harmony between pure mathematics and physics.

NOTES

by

A. SOMMERFELD

The following notes are given in an appendix so as to interfere in no way with Minkowski's text. They are by no means essential, having no other purpose than that of removing certain small formal mathematical difficulties which might hinder the comprehension of Minkowski's great thoughts. The bibliographical references are confined to the literature dealing expressly with the subject of his address. From the physical point of view there is nothing in what Minkowski says that must now be withdrawn, with the exception of the final remark on Newton's law of attraction. What will be the epistemological attitude towards Minkowski's conception of the time-space problem is another question, but, as it seems to me, a question which does not essentially touch his physics.

(1) Page 81, line 8. "On the other hand, the concept of rigid bodies has meaning only in mechanics satisfying the group G_∞." This sentence was confirmed in the widest sense in a discussion on a paper by his disciple M. Born, a year after Minkowski's death. Born (Ann. d. Physik, 30, 1909, p. 1) had defined a relatively rigid body as one in which every element of volume, even in accelerated motions, undergoes the Lorentzian contraction appropriate to its velocity. Ehrenfest (Phys. Zeitschr., 10, 1909, p. 918) showed that such a body cannot be set in rotation; Herglotz (Ann. d. Phys., 31, 1910, p. 393) and F. Nöther (Ann. d. Phys., 31, 1910, p. 919) that it has only three degrees of freedom of movement. The attempt was also made to define a relatively rigid body with six or nine degrees of freedom. But Planck (Phys. Zeitschr., 11, 1910, p. 294) expressed the view that the theory of relativity can operate only with more or less elastic bodies, and Laue (Phys. Zeitschr., 12, 1911, p. 48), employing Minkowski's methods, and his Fig. 2 in the text above, proved that in the theory of relativity every solid body must have an infinite number of degrees of freedom. Finally Herglotz (Ann. d. Physik, 36, 1911, p. 453) developed a relativistic theory of elasticity, according to which elastic tensions always occur if the motion of the body is not relatively rigid in Born's sense. Thus the relatively rigid body plays the same part in this theory of elasticity as the ordinary rigid body plays in the ordinary theory of elasticity.

(2) Page 82, line 18. "If dx/dt for the second band is equal to v, an easy calculation gives $OD' = OC\sqrt{1 - v^2/c^2}$." In Fig. 1, let $\alpha = \angle A'OA$, $\beta = \angle B'OA' = \angle C'OB'$, in which the equality of the last two angles follows from the symmetrical position of the asymptotes with respect to the new axes of co-ordinates (conjugate diameters of the hyperbola).* Since $\alpha + \beta = \frac{1}{4}\pi$,

$$\sin 2\beta = \cos 2\alpha.$$

* Sommerfeld seems to take ct as a co-ordinate in the graph in place of t as used by Minkowski.—TRANS.

In the triangle OD′C′ the law of sines gives

$$\frac{\mathrm{OD'}}{\mathrm{OC'}} = \frac{\sin 2\beta}{\cos \alpha} = \frac{\cos 2\alpha}{\cos \alpha}$$

or, as OC′ = OA′,

$$\mathrm{OD'} = \mathrm{OA'}\frac{\cos 2\alpha}{\cos \alpha} = \mathrm{OA'} \cos \alpha(1 - \tan^2 \alpha) \quad . \quad . \quad . \quad (1)$$

If x, t are the co-ordinates of the point A′ in the x, t system, and therefore x . OA and ct . OC = ct . OA respectively are the corresponding distances from the axes of co-ordinates, we have

$$x\,.\,\mathrm{OA} = \sin \alpha\,.\,\mathrm{OA'}, \quad ct\,.\,\mathrm{OA} = \cos \alpha\,.\,\mathrm{OA'}, \quad \frac{x}{ct} = \tan \alpha = \frac{v}{c} \quad . \quad (2)$$

Inserting these values of x and ct in the equation of the hyperbola, we find

$$\mathrm{OA'}^2(\cos^2 \alpha - \sin^2 \alpha) = \mathrm{OA}^2, \quad \mathrm{OA'} = \frac{\mathrm{OA}}{\cos \alpha \sqrt{(1 - \tan^2 \alpha)}} \quad . \quad (3)$$

therefore, on account of (1) and (2),

$$\mathrm{OD'} = \mathrm{OA}\sqrt{(1 - \tan^2 \alpha)} = \mathrm{OA}\sqrt{(1 - v^2/c^2)}.$$

This, because OA = OC, is the formula to be proved.

Further, in the right-angled triangle OCD,

$$\mathrm{OD} = \frac{\mathrm{OC}}{\cos \alpha} = \frac{\mathrm{OA}}{\cos \alpha}.$$

Equation (3) may therefore be also written in this way,

$$\mathrm{OA'} = \frac{\mathrm{OD}}{\sqrt{(1 - \tan^2 \alpha)}} \text{ or } \frac{\mathrm{OD}}{\mathrm{OA'}} = \sqrt{\left(1 - \frac{v^2}{c^2}\right)}.$$

This, together with (4), gives the proportion,

$$\mathrm{OD} : \mathrm{OA'} = \mathrm{OD'} : \mathrm{OA},$$

which, as OA′ = OC′ and OA = OC, is identical with

$$\mathrm{OD} : \mathrm{OC'} = \mathrm{OD'} : \mathrm{OC}$$

employed on page 82, line 29.

(3) Page 84, line 15. "Any world-point between the front and back cones of O can be arranged, by means of the system of reference, so as to be simultaneous with O, but also just as well so as to be earlier than O, or later than O." M. Laue (Phys. Zeitschr., 12, 1911, p. 48) traces to this observation the proof of Einstein's theorem : In the theory of relativity no process of causality can be propagated with a velocity greater than that of light ("Signal velocity $\leqq c$"). Assume that an event O causes another event P, and that the world-point P lies in the region between the cones of O. In this case the effect would have been conveyed from O to P with a velocity greater than that of light, relatively to the system of reference x, t in question, in which, of course, the effect P is assumed to be later than the cause O, $t_P > 0$. But now, in accordance with the words quoted above, the system of reference may be changed, so that P comes to be earlier than O, that is to say, a system x', t' may be chosen in

infinitely many ways so that t'_P becomes <0. This is irreconcilable with the idea of causality. P must therefore lie either "after" O or on the back cone of O, i.e. the velocity of propagation of a signal to be sent from O, which is to cause a second event at the world-point P, must of necessity be $\leqq c$. (Of course it is possible, even in the theory of relativity, to define processes propagated with velocity greater than light. This can be done geometrically, for example, in a very simple way. But such processes can never serve as signals, i.e. it is impossible to introduce them arbitrarily and by them, for example, to set a relay in motion at a distant place. There may be e.g. optical media, in which the "velocity of light" is greater than c. But in that case what is understood by the velocity of light is the propagation of phases in an infinite periodic wave-train. These can never be used for signalling. On the other hand a wave-front is propagated, in all circumstances and with any constitution of the optical medium, with the velocity c; cf. e.g. A. Sommerfeld, "Festschrift Heinrich Weber," Leipzig, Teubner, 1912, p. 338, or Ann. d. Physik, 44, 1914, p. 177.

(4) Page 85, line 18. As Minkowski once remarked to me, the element of proper time $d\tau$ is not a complete differential. Thus if we connect two world-points O and P by two different world-lines 1 and 2, then

$$\int_1 d\tau \neq \int_2 d\tau.$$

If 1 runs parallel to the t-axis, so that the first transition in the chosen system of reference signifies rest, it is evident that

$$\int_1 d\tau = t, \quad \int_2 d\tau < t.$$

On this depends the retardation of the moving clock compared with the clock at rest. The assertion is based, as Einstein has pointed out, on the unprovable assumption that the clock in motion actually indicates its own proper time, i.e. that it always gives the time corresponding to the state of velocity, regarded as constant, at any instant. The moving clock must naturally have been moved with acceleration (with changes of speed or direction) in order to be compared with the stationary clock at the world-point P. The retardation of the moving clock does not therefore actually indicate "motion," but "accelerated motion." Hence this does not contradict the principle of relativity.

(5) Page 86, line 4. The term "hyperbola of curvature" is formed exactly on the model of the elementary concept of the circle of curvature. The analogy become analytical identity if instead of the real co-ordinate of time t the imaginary $u = ict$ is employed, that is, c times the co-ordinate employed by Minkowski, page 88, line 6.

By page 84 an internal hyperbola in the x, t-plane has the equation, with $k = \rho$,

$$x^2 - c^2t^2 = \rho^2,$$

therefore in the x, u plane

$$x^2 + u^2 = \rho^2.$$

Hence it may be written in parametric form, when ϕ denotes a purely imaginary angle,

$$x = \rho \cos \phi, \; u = \rho \sin \phi.$$

So, as I suggested in the Ann. d. Phys., 33, p. 649, § 8, hyperbolic motion may also be denoted as "cyclic motion," whereby its chief properties (convection of the field, occurrence of a kind of centrifugal force) are characterized with particular clearness. For the hyperbolic motion we have

$$d\tau = \frac{1}{c}\sqrt{(-du^2 - dx^2)} = \frac{\rho}{c}\mid d\phi \mid$$

and thus

$$\dot{x} = \frac{dx}{d\tau} = -ic\sin\phi,\ \dot{u} = \frac{du}{d\tau} = +ic\cos\phi$$

$$\ddot{x} = \frac{d\dot{x}}{d\tau} = \frac{c^2}{\rho}\cos\phi, \quad \ddot{u} = \frac{d\dot{u}}{d\tau} = \frac{c^2}{\rho}\sin\phi.$$

The magnitude of the acceleration vector in hyperbolic motion is therefore c^2/ρ. Since any given world-line is touched by the hyperbola of curvature at three points, it has the same acceleration vector as the hyperbolic motion, and its magnitude is c^2/ρ, as indicated on page 86, line 11.

The centre M of the cyclic motion $x^2 + u^2 = \rho^2$ is evidently the point $x = 0$, $u = 0$, and from this centre all points of the hyperbola have the constant "distance," i.e. a constant magnitude of the radius vector. Therefore ρ denotes the interval marked MP in Fig. 3.

(6) Page 87, line 1. A force X, Y, Z, to be made into a "force vector," must be multiplied by $\dot{t} = dt/d\tau$. This may be explained as follows.

According to Minkowski, page 87, line 10, the momentum vector is defined by $m\dot{x}$, $m\dot{y}$, $m\dot{z}$, $m\dot{t}$, where m denotes the "constant mechanical mass," or, as Minkowski says more plainly elsewhere, the "rest mass." If we retain Newton's law of motion (time rate of change of momentum equal to force), we have to set

$$\frac{d}{dt}(m\dot{x}) = \mathrm{X},\ \frac{d}{dt}(m\dot{y}) = \mathrm{Y},\ \frac{d}{dt}(m\dot{z}) = \mathrm{Z}.$$

Multiplication by $\dot{t}$ makes the left-hand sides into vector components in Minkowski's sense. Therefore $\dot{t}$X, $\dot{t}$Y, $\dot{t}$Z are also the first three components of the "force vector." The fourth component T follows without ambiguity from the requirement that the force vector is to be normal to the motion vector. Minkowski's equations for the mechanics of the mass point are therefore, with constant rest mass,

$$m\ddot{x} = \dot{t}\mathrm{X},\ \ m\ddot{y} = \dot{t}\mathrm{Y},\ \ m\ddot{z} = \dot{t}\mathrm{Z},\ \ m\ddot{t} = \dot{t}\mathrm{T}.$$

The assumption of constancy of rest mass can only be maintained, however, when the energy-content of the body is not changed in its motion, or in the words of Planck, when the motion ensues "adiabatically and isochorically."

(7) Pages 88 and 89. What is characteristic of the constructions here given, is their complete independence of any special system of reference. They give, as Minkowski postulates on page 88, "reciprocal relations between world-lines" (or world-points) as "the most perfect expression of physical laws." On page 89, for example, the electrodynamic potential (four-potential) is not referred to the axes of co-ordinates x, y, z, t until it is to be conventionally divided into a scalar and a vector portion, which have no independent invariant meaning from the relativistic standpoint.

By way of commentary to Minkowski I have deduced, from Maxwell's equations, by Minkowski's methods, an invariant analytical form for the four-potential and the ponderomotive action between two electrons, and so given another view of these constructions of Minkowski. Instead of going into details here, I may refer to my article in Ann. d. Phys., 33, 1910, p. 649, § 7, or to M. Laue, "Das Relativitätsprinzip," Braunschweig, Vieweg, 1913, § 19. Compare also Minkowski's address on the principle of relativity, edited by myself, in Ann. d. Phys., 47, 1915, p. 927, where the four-potential is placed at the head of electrodynamics, and this theory thus reduced to its simplest form.

(8) Page 89, line 6. The invariant representation of the electromagnetic field by a "vector of the second kind" (or, as I proposed to call it, a "six-vector," a term which seems to be winning acceptance) is a particularly important part of Minkowski's view of electrodynamics. Whereas Minkowski's ideas on the vector of the first kind, or four-vector, were in part anticipated by Poincaré (Rend. Circ. Mat. Palermo, 21, 1906), the introduction of the six-vector is new. Like the six-vector, the wrench of mechanics (standing for a single force and a couple) depends on six independent parameters. And as in the electromagnetic field "the separation into electric and magnetic force is a relative one," so with the wrench, as is well known, the division into single force and couple can be made in very many ways.

(9) Page 90, line 9. Minkowski's relativistic form of Newton's law for the special case of zero acceleration mentioned in the text is included in the more general form proposed by Poincaré (loc. cit.). On the other hand, in taking acceleration into consideration, it goes further than the latter. Minkowski's or Poincaré's formulation of the law of gravitation shows that it is possible in many ways to reconcile Newton's law with the theory of relativity. That law is viewed as a point law, and gravitation therefore in a certain sense as action at a distance. The general theory of relativity, which Einstein has been developing from 1907 on, gets a deeper grip of the problem of gravitation. Gravitation is not only regarded as a field action and described by space-time differential equations—which seems from the present standpoint irrefutable—but it is also united organically with the principle of relativity extended to any transformations, whereas Minkowski and Poincaré had adapted it to the postulate of relativity in a more external manner. In the general theory of relativity the space-time structure is determined, from or together with, gravitation. Thus the principle of relativity, by an extension of Minkowski's ideas, is so formulated that it postulates the co-variance of physical quantities with reference to all point transformations, so that the coefficients of the invariant linear element enter into the laws of physics.

(10) Page 90, line 33. The "fundamental equations for electromagnetic processes in ponderable bodies" are developed by Minkowski in Göttinger Nachrichten, 1907. It was not granted him to complete the "deduction of this equation on the basis of the theory of electrons." His essays in this direction have been worked out by M. Born, and together with the "Fundamental Equations" make up the first volume of the series of monographs edited by Otto Blumenthal (Leipzig, 1910).

ON THE INFLUENCE OF GRAVITATION ON THE PROPAGATION OF LIGHT

BY

A. EINSTEIN

Translated from "Über den Einfluss der Schwerkraft auf die Ausbreitung des Lichtes," Annalen der Physik, 35, 1911.

ON THE INFLUENCE OF GRAVITATION ON THE PROPAGATION OF LIGHT

By A. EINSTEIN

IN a memoir published four years ago * I tried to answer the question whether the propagation of light is influenced by gravitation. I return to this theme, because my previous presentation of the subject does not satisfy me, and for a stronger reason, because I now see that one of the most important consequences of my former treatment is capable of being tested experimentally. For it follows from the theory here to be brought forward, that rays of light, passing close to the sun, are deflected by its gravitational field, so that the angular distance between the sun and a fixed star appearing near to it is apparently increased by nearly a second of arc.

In the course of these reflexions further results are yielded which relate to gravitation. But as the exposition of the entire group of considerations would be rather difficult to follow, only a few quite elementary reflexions will be given in the following pages, from which the reader will readily be able to inform himself as to the suppositions of the theory and its line of thought. The relations here deduced, even if the theoretical foundation is sound, are valid only to a first approximation.

§ 1. A Hypothesis as to the Physical Nature of the Gravitational Field

In a homogeneous gravitational field (acceleration of gravity γ) let there be a stationary system of co-ordinates K, orientated so that the lines of force of the gravitational field run in the negative direction of the axis of z. In a space free

* A. Einstein, Jahrbuch für Radioakt. und Elektronik, 4, 1907.

of gravitational fields let there be a second system of co-ordinates K′, moving with uniform acceleration (γ) in the positive direction of its axis of z. To avoid unnecessary complications, let us for the present disregard the theory of relativity, and regard both systems from the customary point of view of kinematics, and the movements occurring in them from that of ordinary mechanics.

Relatively to K, as well as relatively to K′, material points which are not subjected to the action of other material points, move in keeping with the equations

$$\frac{d^2x}{dt^2} = 0, \frac{d^2y}{dt^2} = 0, \frac{d^2z}{dt^2} = -\gamma.$$

For the accelerated system K′ this follows directly from Galileo's principle, but for the system K, at rest in a homogeneous gravitational field, from the experience that all bodies in such a field are equally and uniformly accelerated. This experience, of the equal falling of all bodies in the gravitational field, is one of the most universal which the observation of nature has yielded; but in spite of that the law has not found any place in the foundations of our edifice of the physical universe.

But we arrive at a very satisfactory interpretation of this law of experience, if we assume that the systems K and K′ are physically exactly equivalent, that is, if we assume that we may just as well regard the system K as being in a space free from gravitational fields, if we then regard K as uniformly accelerated. This assumption of exact physical equivalence makes it impossible for us to speak of the absolute acceleration of the system of reference, just as the usual theory of relativity forbids us to talk of the absolute velocity of a system; * and it makes the equal falling of all bodies in a gravitational field seem a matter of course.

As long as we restrict ourselves to purely mechanical processes in the realm where Newton's mechanics holds sway, we are certain of the equivalence of the systems K and K′.

* Of course we cannot replace any arbitrary gravitational field by a state of motion of the system without a gravitational field, any more than, by a transformation of relativity, we can transform all points of a medium in any kind of motion to rest.

But this view of ours will not have any deeper significance unless the systems K and K′ are equivalent with respect to all physical processes, that is, unless the laws of nature with respect to K are in entire agreement with those with respect to K′. By assuming this to be so, we arrive at a principle which, if it is really true, has great heuristic importance. For by theoretical consideration of processes which take place relatively to a system of reference with uniform acceleration, we obtain information as to the career of processes in a homogeneous gravitational field. We shall now show, first of all, from the standpoint of the ordinary theory of relativity, what degree of probability is inherent in our hypothesis.

§ 2. On the Gravitation of Energy

One result yielded by the theory of relativity is that the inertia mass of a body increases with the energy it contains; if the increase of energy amounts to E, the increase in inertia mass is equal to E/c^2, when c denotes the velocity of light. Now is there an increase of gravitating mass corresponding to this increase of inertia mass? If not, then a body would fall in the same gravitational field with varying acceleration according to the energy it contained. That highly satisfactory result of the theory of relativity by which the law of the conservation of mass is merged in the law of conservation of energy could not be maintained, because it would compel us to abandon the law of the conservation of mass in its old form for inertia mass, and maintain it for gravitating mass.

But this must be regarded as very improbable. On the other hand, the usual theory of relativity does not provide us with any argument from which to infer that the weight of a body depends on the energy contained in it. But we shall show that our hypothesis of the equivalence of the systems K and K′ gives us gravitation of energy as a necessary consequence.

Let the two material systems S_1 and S_2, provided with instruments of measurement, be situated on the z-axis of K at the distance h from each other,* so that the gravitation potential in S_2 is greater than that in S_1 by γh. Let a definite quantity

* The dimensions of S_1 and S_2 are regarded as infinitely small in comparison with h.

of energy E be emitted from S_2 towards S_1. Let the quantities of energy in S_1 and S_2 be measured by contrivances which—brought to one place in the system z and there compared—shall be perfectly alike. As to the process of this conveyance of energy by radiation we can make no *a priori* assertion, because we do not know the influence of the gravitational field on the radiation and the measuring instruments in S_1 and S_2.

But by our postulate of the equivalence of K and K′ we are able, in place of the system K in a homogeneous gravitational field, to set the gravitation-free system K′, which moves with uniform acceleration in the direction of positive z, and with the z-axis of which the material systems S_1 and S_2 are rigidly connected.

We judge of the process of the transference of energy by radiation from S_2 to S_1 from a system K_0, which is to be free from acceleration. At the moment when the radiation energy E_2 is emitted from S_2 toward S_1, let the velocity of K′ relatively to K_0 be zero. The radiation will arrive at S_1 when the time h/c has elapsed (to a first approximation). But at this moment the velocity of S_1 relatively to K_0 is $\gamma h/c = v$. Therefore by the ordinary theory of relativity the radiation arriving at S_1 does not possess the energy E_2, but a greater energy E_1, which is related to E_2 to a first approximation by the equation *

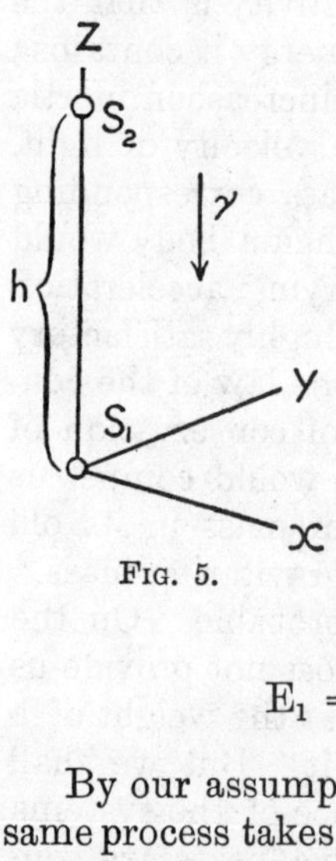

FIG. 5.

$$E_1 = E_2\left(1 + \frac{v}{c}\right) = E_2\left(1 + \gamma\frac{h}{c^2}\right) \quad . \quad . \quad (1)$$

By our assumption exactly the same relation holds if the same process takes place in the system K, which is not accelerated, but is provided with a gravitational field. In this case we may replace γh by the potential Φ of the gravitation vector in S_2, if the arbitrary constant of Φ in S_1 is equated to zero. We then have the equation

$$E_1 = E_2 + \frac{E_2}{c^2}\Phi \quad . \quad . \quad . \quad (1a)$$

* See above, pp. 69-71.

This equation expresses the law of energy for the process under observation. The energy E_1 arriving at S_1 is greater than the energy E_2, measured by the same means, which was emitted in S_2, the excess being the potential energy of the mass E_2/c^2 in the gravitational field. It thus proves that for the fulfilment of the principle of energy we have to ascribe to the energy E, before its emission in S_2, a potential energy due to gravity, which corresponds to the gravitational mass E/c^2. Our assumption of the equivalence of K and K′ thus removes the difficulty mentioned at the beginning of this paragraph which is left unsolved by the ordinary theory of relativity.

The meaning of this result is shown particularly clearly if we consider the following cycle of operations :—

1. The energy E, as measured in S_2, is emitted in the form of radiation in S_2 towards S_1, where, by the result just obtained, the energy $E(1 + \gamma h/c^2)$, as measured in S_1, is absorbed.

2. A body W of mass M is lowered from S_2 to S_1, work $M\gamma h$ being done in the process.

3. The energy E is transferred from S_1 to the body W while W is in S_1. Let the gravitational mass M be thereby changed so that it acquires the value M′.

4. Let W be again raised to S_2, work $M'\gamma h$ being done in the process.

5. Let E be transferred from W back to S_2.

The effect of this cycle is simply that S_1 has undergone the increase of energy $E\gamma h/c^2$, and that the quantity of energy $M'\gamma h - M\gamma h$ has been conveyed to the system in the form of mechanical work. By the principle of energy, we must therefore have

$$E\gamma\frac{h}{c^2} = M'\gamma h - M\gamma h,$$

or

$$M' - M = E/c^2. \qquad (1b)$$

The increase in gravitational mass is thus equal to E/c^2, and therefore equal to the increase in inertia mass as given by the theory of relativity.

The result emerges still more directly from the equivalence

of the systems K and K′, according to which the gravitational mass in respect of K is exactly equal to the inertia mass in respect of K′; energy must therefore possess a gravitational mass which is equal to its inertia mass. If a mass M_0 be suspended on a spring balance in the system K′, the balance will indicate the apparent weight $M_0\gamma$ on account of the inertia of M_0. If the quantity of energy E be transferred to M_0, the spring balance, by the law of the inertia of energy, will indicate $(M_0 + E/c^2)\gamma$. By reason of our fundamental assumption exactly the same thing must occur when the experiment is repeated in the system K, that is, in the gravitational field.

§ 3. Time and the Velocity of Light in the Gravitational Field

If the radiation emitted in the uniformly accelerated system K′ in S_2 toward S_1 had the frequency ν_2 relatively to the clock in S_2, then, relatively to S_1, at its arrival in S_1 it no longer has the frequency ν_2 relatively to an identical clock in S_1, but a greater frequency ν_1, such that to a first approximation

$$\nu_1 = \nu_2\left(1 + \gamma\frac{h}{c^2}\right) \quad . \quad . \quad . \quad (2)$$

For if we again introduce the unaccelerated system of reference K_0, relatively to which, at the time of the emission of light, K′ has no velocity, then S_1, at the time of arrival of the radiation at S_1, has, relatively to K_0, the velocity $\gamma h/c$, from which, by Doppler's principle, the relation as given results immediately.

In agreement with our assumption of the equivalence of the systems K′ and K, this equation also holds for the stationary system of co-ordinates K, provided with a uniform gravitational field, if in it the transference by radiation takes place as described. It follows, then, that a ray of light emitted in S_2 with a definite gravitational potential, and possessing at its emission the frequency ν_2—compared with a clock in S_2—will, at its arrival in S_1, possess a different frequency ν_1—measured by an identical clock in S_1. For γh we substitute the gravitational potential Φ of S_2—that of S_1

being taken as zero—and assume that the relation which we have deduced for the homogeneous gravitational field also holds for other forms of field. Then

$$\nu_1 = \nu_2\left(1 + \frac{\Phi}{c^2}\right) \quad . \quad . \quad . \quad . \quad (2a)$$

This result (which by our deduction is valid to a first approximation) permits, in the first place, of the following application. Let ν_0 be the vibration-number of an elementary light-generator, measured by a delicate clock at the same place. Let us imagine them both at a place on the surface of the Sun (where our S_2 is located). Of the light there emitted, a portion reaches the Earth (S_1), where we measure the frequency of the arriving light with a clock U in all respects resembling the one just mentioned. Then by (2a),

$$\nu = \nu_0\left(1 + \frac{\Phi}{c^2}\right),$$

where Φ is the (negative) difference of gravitational potential between the surface of the Sun and the Earth. Thus according to our view the spectral lines of sunlight, as compared with the corresponding spectral lines of terrestrial sources of light, must be somewhat displaced toward the red, in fact by the relative amount

$$\frac{\nu_0 - \nu}{\nu_0} = -\frac{\Phi}{c^2} = 2 \,.\, 10^{-6}$$

If the conditions under which the solar bands arise were exactly known, this shifting would be susceptible of measurement. But as other influences (pressure, temperature) affect the position of the centres of the spectral lines, it is difficult to discover whether the inferred influence of the gravitational potential really exists.*

On a superficial consideration equation (2), or (2a), respectively, seems to assert an absurdity. If there is constant transmission of light from S_2 to S_1, how can any other number of periods per second arrive in S_1 than is emitted

* L. F. Jewell (Journ. de Phys., 6, 1897, p. 84) and particularly Ch. Fabry and H. Boisson (Comptes rendus, 148, 1909, pp. 688-690) have actually found such displacements of fine spectral lines toward the red end of the spectrum, of the order of magnitude here calculated, but have ascribed them to an effect of pressure in the absorbing layer.

in S_2? But the answer is simple. We cannot regard ν_2 or respectively ν_1 simply as frequencies (as the number of periods per second) since we have not yet determined the time in system K. What ν_2 denotes is the number of periods with reference to the time-unit of the clock U in S_2, while ν_1 denotes the number of periods per second with reference to the identical clock in S_1. Nothing compels us to assume that the clocks U in different gravitation potentials must be regarded as going at the same rate. On the contrary, we must certainly define the time in K in such a way that the number of wave crests and troughs between S_2 and S_1 is independent of the absolute value of time; for the process under observation is by nature a stationary one. If we did not satisfy this condition, we should arrive at a definition of time by the application of which time would merge explicitly into the laws of nature, and this would certainly be unnatural and unpractical. Therefore the two clocks in S_1 and S_2 do not both give the "time" correctly. If we measure time in S_1 with the clock U, then we must measure time in S_2 with a clock which goes $1 + \Phi/c^2$ times more slowly than the clock U when compared with U at one and the same place. For when measured by such a clock the frequency of the ray of light which is considered above is at its emission in S_2

$$\nu_2\left(1 + \frac{\Phi}{c^2}\right)$$

and is therefore, by (2a), equal to the frequency ν_1 of the same ray of light on its arrival in S_1.

This has a consequence which is of fundamental importance for our theory. For if we measure the velocity of light at different places in the accelerated, gravitation-free system K′, employing clocks U of identical constitution, we obtain the same magnitude at all these places. The same holds good, by our fundamental assumption, for the system K as well. But from what has just been said we must use clocks of unlike constitution, for measuring time at places with differing gravitation potential. For measuring time at a place which, relatively to the origin of the co-ordinates, has the gravitation potential Φ, we must employ a clock which—when removed to the origin of co-ordinates—goes $(1 + \Phi/c^2)$ times more slowly than the clock used for measuring time at the origin of co-ordinates. If we call the velocity of light at the origin of co-ordinates c_0, then the velocity of light c at a place with the gravitation potential Φ will be given by the relation

$$c = c_0\left(1 + \frac{\Phi}{c^2}\right) \quad . \quad . \quad . \quad . \quad (3)$$

The principle of the constancy of the velocity of light holds good according to this theory in a different form from that which usually underlies the ordinary theory of relativity.

§ 4. Bending of Light-Rays in the Gravitational Field

From the proposition which has just been proved, that the velocity of light in the gravitational field is a function of the place, we may easily infer, by means of Huyghens's principle, that light-rays propagated across a gravitational field undergo deflexion. For let E be a wave front of a plane light-wave at the time t, and let P_1 and P_2 be two points in that plane at

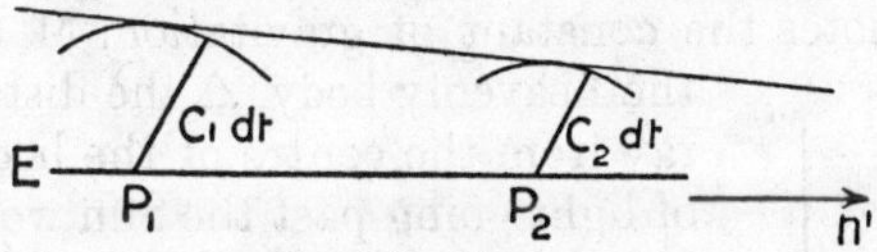

FIG. 6.

unit distance from each other. P_1 and P_2 lie in the plane of the paper, which is chosen so that the differential coefficient of Φ, taken in the direction of the normal to the plane, vanishes, and therefore also that of c. We obtain the corresponding wave front at time $t + dt$, or, rather, its line of section with the plane of the paper, by describing circles round the points P_1 and P_2 with radii $c_1 dt$ and $c_2 dt$ respectively, where c_1 and c_2 denote the velocity of light at the points P_1 and P_2 respectively, and by drawing the tangent to these circles. The angle through which the light-ray is deflected in the path cdt is therefore

$$(c_1 - c_2)dt = -\frac{\partial c}{\partial n'}dt,$$

if we calculate the angle positively when the ray is bent toward the side of increasing n'. The angle of deflexion per unit of path of the light-ray is thus

$$-\frac{1}{c}\frac{\partial c}{\partial n'}, \text{ or by (3) } -\frac{1}{c^2}\frac{\partial \Phi}{\partial n'}.$$

Finally, we obtain for the deflexion which a light-ray experiences toward the side n' on any path (s) the expression

$$a = -\frac{1}{c^2}\int\frac{\partial\Phi}{\partial n'}ds \quad . \quad . \quad . \quad . \quad (4)$$

We might have obtained the same result by directly considering the propagation of a ray of light in the uniformly accelerated system K′, and transferring the result to the system K, and thence to the case of a gravitational field of any form.

By equation (4) a ray of light passing along by a heavenly body suffers a deflexion to the side of the diminishing gravitational potential, that is, on the side directed toward the heavenly body, of the magnitude

$$a = \frac{1}{c^2}\int_{\theta=-\frac{1}{2}\pi}^{\theta=\frac{1}{2}\pi}\frac{kM}{r^2}\cos\theta ds = 2\frac{kM}{c^2\Delta}$$

where k denotes the constant of gravitation, M the mass of the heavenly body, Δ the distance of the ray from the centre of the body. A ray of light going past the Sun would accordingly undergo deflexion to the amount of $4 \cdot 10^{-6} = \cdot 83$ seconds of arc. The angular distance of the star from the centre of the Sun appears to be increased by this amount. As the fixed stars in the parts of the sky near the Sun are visible during total eclipses of the Sun, this consequence of the theory may be compared with experience. With the planet Jupiter the displacement to be expected reaches to about $\frac{1}{100}$ of the amount given. It would be a most desirable thing if astronomers would take up the question here raised. For apart from any theory there is the question whether it is possible with the equipment at present available to detect an influence of gravitational fields on the propagation of light.

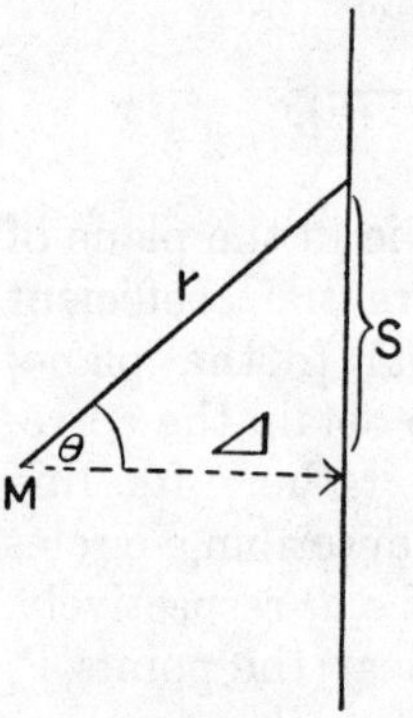

FIG. 7.

THE FOUNDATION OF THE GENERAL THEORY OF RELATIVITY

BY

A. EINSTEIN

Translated from "Die Grundlage der allgemeinen Relativitätstheorie," Annalen der Physik, 49, 1916.

THE FOUNDATION OF THE GENERAL THEORY OF RELATIVITY

By A. EINSTEIN

A. Fundamental Considerations on the Postulate of Relativity

§ 1. Observations on the Special Theory of Relativity

THE special theory of relativity is based on the following postulate, which is also satisfied by the mechanics of Galileo and Newton.

If a system of co-ordinates K is chosen so that, in relation to it, physical laws hold good in their simplest form, the *same* laws also hold good in relation to any other system of co-ordinates K′ moving in uniform translation relatively to K. This postulate we call the "special principle of relativity." The word "special" is meant to intimate that the principle is restricted to the case when K′ has a motion of uniform translation relatively to K, but that the equivalence of K′ and K does not extend to the case of non-uniform motion of K′ relatively to K.

Thus the special theory of relativity does not depart from classical mechanics through the postulate of relativity, but through the postulate of the constancy of the velocity of light *in vacuo*, from which, in combination with the special principle of relativity, there follow, in the well-known way, the relativity of simultaneity, the Lorentzian transformation, and the related laws for the behaviour of moving bodies and clocks.

The modification to which the special theory of relativity has subjected the theory of space and time is indeed far-reaching, but one important point has remained unaffected.

For the laws of geometry, even according to the special theory of relativity, are to be interpreted directly as laws relating to the possible relative positions of solid bodies at rest; and, in a more general way, the laws of kinematics are to be interpreted as laws which describe the relations of measuring bodies and clocks. To two selected material points of a stationary rigid body there always corresponds a distance of quite definite length, which is independent of the locality and orientation of the body, and is also independent of the time. To two selected positions of the hands of a clock at rest relatively to the privileged system of reference there always corresponds an interval of time of a definite length, which is independent of place and time. We shall soon see that the general theory of relativity cannot adhere to this simple physical interpretation of space and time.

§ 2. The Need for an Extension of the Postulate of Relativity

In classical mechanics, and no less in the special theory of relativity, there is an inherent epistemological defect which was, perhaps for the first time, clearly pointed out by Ernst Mach. We will elucidate it by the following example:—Two fluid bodies of the same size and nature hover freely in space at so great a distance from each other and from all other masses that only those gravitational forces need be taken into account which arise from the interaction of different parts of the same body. Let the distance between the two bodies be invariable, and in neither of the bodies let there be any relative movements of the parts with respect to one another. But let either mass, as judged by an observer at rest relatively to the other mass, rotate with constant angular velocity about the line joining the masses. This is a verifiable relative motion of the two bodies. Now let us imagine that each of the bodies has been surveyed by means of measuring instruments at rest relatively to itself, and let the surface of S_1 prove to be a sphere, and that of S_2 an ellipsoid of revolution. Thereupon we put the question—What is the reason for this difference in the two bodies? No answer can be admitted as epistemologically satisfactory,* unless the reason given is an *observable fact of experience.* The law of causality has not the significance of a statement as to the world of experience, except when *observable facts* ultimately appear as causes and effects.

Newtonian mechanics does not give a satisfactory answer to this question. It pronounces as follows:—The laws of mechanics apply to the space R_1, in respect to which the body S_1 is at rest, but not to the space R_2, in respect to which the body S_2 is at rest. But the privileged space R_1 of Galileo, thus introduced, is a merely *factitious* cause, and not a thing that can be observed. It is therefore clear that Newton's mechanics does not really satisfy the requirement of causality in the case under consideration, but only apparently does so, since it makes the factitious cause R_1 responsible for the observable difference in the bodies S_1 and S_2.

The only satisfactory answer must be that the physical system consisting of S_1 and S_2 reveals within itself no imaginable cause to which the differing behaviour of S_1 and S_2 can be referred. The cause must therefore lie *outside* this system. We have to take it that the general laws of motion, which in particular determine the shapes of S_1 and S_2, must be such that the mechanical behaviour of S_1 and S_2 is partly conditioned, in quite essential respects, by distant masses which we have not included in the system under consideration. These distant masses and their motions relative to S_1 and S_2 must then be regarded as the seat of the causes (which must be susceptible to observation) of the different behaviour of our two bodies S_1 and S_2. They take over the rôle of the factitious cause R_1. Of all imaginable spaces R_1, R_2, etc., in any kind of motion relatively to one another, there is none which we may look upon as privileged *a priori* without reviving the above-mentioned epistemological objection. *The laws of physics must be of such a nature that they apply to systems of reference in any kind of motion.* Along this road we arrive at an extension of the postulate of relativity.

In addition to this weighty argument from the theory of

* Of course an answer may be satisfactory from the point of view of epistemology, and yet be unsound physically, if it is in conflict with other experiences.

knowledge, there is a well-known physical fact which favours an extension of the theory of relativity. Let K be a Galilean system of reference, i.e. a system relatively to which (at least in the four-dimensional region under consideration) a mass, sufficiently distant from other masses, is moving with uniform motion in a straight line. Let K′ be a second system of reference which is moving relatively to K in *uniformly accelerated* translation. Then, relatively to K′, a mass sufficiently distant from other masses would have an accelerated motion such that its acceleration and direction of acceleration are independent of the material composition and physical state of the mass.

Does this permit an observer at rest relatively to K′ to infer that he is on a "really" accelerated system of reference? The answer is in the negative; for the above-mentioned relation of freely movable masses to K′ may be interpreted equally well in the following way. The system of reference K′ is unaccelerated, but the space-time territory in question is under the sway of a gravitational field, which generates the accelerated motion of the bodies relatively to K′.

This view is made possible for us by the teaching of experience as to the existence of a field of force, namely, the gravitational field, which possesses the remarkable property of imparting the same acceleration to all bodies.* The mechanical behaviour of bodies relatively to K′ is the same as presents itself to experience in the case of systems which we are wont to regard as "stationary" or as "privileged." Therefore, from the physical standpoint, the assumption readily suggests itself that the systems K and K′ may both with equal right be looked upon as "stationary," that is to say, they have an equal title as systems of reference for the physical description of phenomena.

It will be seen from these reflexions that in pursuing the general theory of relativity we shall be led to a theory of gravitation, since we are able to "produce" a gravitational field merely by changing the system of co-ordinates. It will also be obvious that the principle of the constancy of the velocity of light *in vacuo* must be modified, since we easily

* Eötvös has proved experimentally that the gravitational field has this property in great accuracy.

recognize that the path of a ray of light with respect to K′ must in general be curvilinear, if with respect to K light is propagated in a straight line with a definite constant velocity.

§ 3. The Space-Time Continuum. Requirement of General Co-Variance for the Equations Expressing General Laws of Nature

In classical mechanics, as well as in the special theory of relativity, the co-ordinates of space and time have a direct physical meaning. To say that a point-event has the X_1 co-ordinate x_1 means that the projection of the point-event on the axis of X_1, determined by rigid rods and in accordance with the rules of Euclidean geometry, is obtained by measuring off a given rod (the unit of length) x_1 times from the origin of co-ordinates along the axis of X_1. To say that a point-event has the X_4 co-ordinate $x_4 = t$, means that a standard clock, made to measure time in a definite unit period, and which is stationary relatively to the system of co-ordinates and practically coincident in space with the point-event,* will have measured off $x_4 = t$ periods at the occurrence of the event.

This view of space and time has always been in the minds of physicists, even if, as a rule, they have been unconscious of it. This is clear from the part which these concepts play in physical measurements; it must also have underlain the reader's reflexions on the preceding paragraph (§ 2) for him to connect any meaning with what he there read. But we shall now show that we must put it aside and replace it by a more general view, in order to be able to carry through the postulate of general relativity, if the special theory of relativity applies to the special case of the absence of a gravitational field.

In a space which is free of gravitational fields we introduce a Galilean system of reference K (x, y, z, t), and also a system of co-ordinates K′ (x', y', z', t') in uniform rotation relatively to K. Let the origins of both systems, as well as their axes

* We assume the possibility of verifying "simultaneity" for events immediately proximate in space, or—to speak more precisely—for immediate proximity or coincidence in space-time, without giving a definition of this fundamental concept.

of Z, permanently coincide. We shall show that for a space-time measurement in the system K′ the above definition of the physical meaning of lengths and times cannot be maintained. For reasons of symmetry it is clear that a circle around the origin in the X, Y plane of K may at the same time be regarded as a circle in the X′, Y′ plane of K′. We suppose that the circumference and diameter of this circle have been measured with a unit measure infinitely small compared with the radius, and that we have the quotient of the two results. If this experiment were performed with a measuring-rod at rest relatively to the Galilean system K, the quotient would be π. With a measuring-rod at rest relatively to K′, the quotient would be greater than π. This is readily understood if we envisage the whole process of measuring from the "stationary" system K, and take into consideration that the measuring-rod applied to the periphery undergoes a Lorentzian contraction, while the one applied along the radius does not. Hence Euclidean geometry does not apply to K′. The notion of co-ordinates defined above, which presupposes the validity of Euclidean geometry, therefore breaks down in relation to the system K′. So, too, we are unable to introduce a time corresponding to physical requirements in K′, indicated by clocks at rest relatively to K′. To convince ourselves of this impossibility, let us imagine two clocks of identical constitution placed, one at the origin of co-ordinates, and the other at the circumference of the circle, and both envisaged from the "stationary" system K. By a familiar result of the special theory of relativity, the clock at the circumference—judged from K—goes more slowly than the other, because the former is in motion and the latter at rest. An observer at the common origin of co-ordinates, capable of observing the clock at the circumference by means of light, would therefore see it lagging behind the clock beside him. As he will not make up his mind to let the velocity of light along the path in question depend explicitly on the time, he will interpret his observations as showing that the clock at the circumference "really" goes more slowly than the clock at the origin. So he will be obliged to define time in such a way that the rate of a clock depends upon where the clock may be.

We therefore reach this result:—In the general theory of relativity, space and time cannot be defined in such a way that differences of the spatial co-ordinates can be directly measured by the unit measuring-rod, or differences in the time co-ordinate by a standard clock.

The method hitherto employed for laying co-ordinates into the space-time continuum in a definite manner thus breaks down, and there seems to be no other way which would allow us to adapt systems of co-ordinates to the four-dimensional universe so that we might expect from their application a particularly simple formulation of the laws of nature. So there is nothing for it but to regard all imaginable systems of co-ordinates, on principle, as equally suitable for the description of nature. This comes to requiring that:—

The general laws of nature are to be expressed by equations which hold good for all systems of co-ordinates, that is, are co-variant with respect to any substitutions whatever (generally co-variant).

It is clear that a physical theory which satisfies this postulate will also be suitable for the general postulate of relativity. For the sum of *all* substitutions in any case includes those which correspond to all relative motions of three-dimensional systems of co-ordinates. That this requirement of general co-variance, which takes away from space and time the last remnant of physical objectivity, is a natural one, will be seen from the following reflexion. All our space-time verifications invariably amount to a determination of space-time coincidences. If, for example, events consisted merely in the motion of material points, then ultimately nothing would be observable but the meetings of two or more of these points. Moreover, the results of our measurings are nothing but verifications of such meetings of the material points of our measuring instruments with other material points, coincidences between the hands of a clock and points on the clock dial, and observed point-events happening at the same place at the same time.

The introduction of a system of reference serves no other purpose than to facilitate the description of the totality of such coincidences. We allot to the universe four space-time variables x_1, x_2, x_3, x_4 in such a way that for every point-event there is a corresponding system of values of the variables $x_1 \ldots x_4$. To two coincident point-events there corresponds one system of values of the variables $x_1 \ldots x_4$, i.e. coincidence is characterized by the identity of the co-ordinates. If, in place of the variables $x_1 \ldots x_4$, we introduce functions of them, x'_1, x'_2, x'_3, x'_4, as a new system of co-ordinates, so that the systems of values are made to correspond to one another without ambiguity, the equality of all four co-ordinates in the new system will also serve as an expression for the space-time coincidence of the two point-events. As all our physical experience can be ultimately reduced to such coincidences, there is no immediate reason for preferring certain systems of co-ordinates to others, that is to say, we arrive at the requirement of general co-variance.

§ 4. The Relation of the Four Co-ordinates to Measurement in Space and Time

It is not my purpose in this discussion to represent the general theory of relativity as a system that is as simple and logical as possible, and with the minimum number of axioms; but my main object is to develop this theory in such a way that the reader will feel that the path we have entered upon is psychologically the natural one, and that the underlying assumptions will seem to have the highest possible degree of security. With this aim in view let it now be granted that:—

For infinitely small four-dimensional regions the theory of relativity in the restricted sense is appropriate, if the co-ordinates are suitably chosen.

For this purpose we must choose the acceleration of the infinitely small ("local") system of co-ordinates so that no gravitational field occurs; this is possible for an infinitely small region. Let X_1, X_2, X_3, be the co-ordinates of space, and X_4 the appertaining co-ordinate of time measured in the appropriate unit.* If a rigid rod is imagined to be given as the unit measure, the co-ordinates, with a given orientation of the system of co-ordinates, have a direct physical meaning

* The unit of time is to be chosen so that the velocity of light *in vacuo* as measured in the "local" system of co-ordinates is to be equal to unity.

in the sense of the special theory of relativity. By the special theory of relativity the expression

$$ds^2 = -dX_1^2 - dX_2^2 - dX_3^2 + dX_4^2 \quad . \quad . \quad (1)$$

then has a value which is independent of the orientation of the local system of co-ordinates, and is ascertainable by measurements of space and time. The magnitude of the linear element pertaining to points of the four-dimensional continuum in infinite proximity, we call ds. If the ds belonging to the element $dX_1 \ldots dX_4$ is positive, we follow Minkowski in calling it time-like; if it is negative, we call it space-like.

To the "linear element" in question, or to the two infinitely proximate point-events, there will also correspond definite differentials $dx_1 \ldots dx_4$ of the four-dimensional co-ordinates of any chosen system of reference. If this system, as well as the "local" system, is given for the region under consideration, the dX_ν will allow themselves to be represented here by definite linear homogeneous expressions of the dx_σ:—

$$dX_\nu = \sum_\sigma a_{\nu\sigma} dx_\sigma \quad . \quad . \quad . \quad . \quad (2)$$

Inserting these expressions in (1), we obtain

$$ds^2 = \sum_{\tau\sigma} g_{\sigma\tau} dx_\sigma dx_\tau, \quad . \quad . \quad . \quad . \quad (3)$$

where the $g_{\sigma\tau}$ will be functions of the x_σ. These can no longer be dependent on the orientation and the state of motion of the "local" system of co-ordinates, for ds^2 is a quantity ascertainable by rod-clock measurement of point-events infinitely proximate in space-time, and defined independently of any particular choice of co-ordinates. The $g_{\sigma\tau}$ are to be chosen here so that $g_{\sigma\tau} = g_{\tau\sigma}$; the summation is to extend over all values of σ and τ, so that the sum consists of 4×4 terms, of which twelve are equal in pairs.

The case of the ordinary theory of relativity arises out of the case here considered, if it is possible, by reason of the particular relations of the $g_{\sigma\tau}$ in a finite region, to choose the system of reference in the finite region in such a way that the $g_{\sigma\tau}$ assume the constant values

$$\left.\begin{matrix} -1 & 0 & 0 & 0 \\ 0 & -1 & 0 & 0 \\ 0 & 0 & -1 & 0 \\ 0 & 0 & 0 & +1 \end{matrix}\right\} \quad . \quad . \quad . \quad (4)$$

We shall find hereafter that the choice of such co-ordinates is, in general, not possible for a finite region.

From the considerations of § 2 and § 3 it follows that the quantities $g_{\tau\sigma}$ are to be regarded from the physical standpoint as the quantities which describe the gravitational field in relation to the chosen system of reference. For, if we now assume the special theory of relativity to apply to a certain four-dimensional region with the co-ordinates properly chosen, then the $g_{\sigma\tau}$ have the values given in (4). A free material point then moves, relatively to this system, with uniform motion in a straight line. Then if we introduce new space-time co-ordinates x_1, x_2, x_3, x_4, by means of any substitution we choose, the $g^{\sigma\tau}$ in this new system will no longer be constants, but functions of space and time. At the same time the motion of the free material point will present itself in the new co-ordinates as a curvilinear non-uniform motion, and the law of this motion will be independent of the nature of the moving particle. We shall therefore interpret this motion as a motion under the influence of a gravitational field. We thus find the occurrence of a gravitational field connected with a space-time variability of the g_{σ} . So, too, in the general case, when we are no longer able by a suitable choice of co-ordinates to apply the special theory of relativity to a finite region, we shall hold fast to the view that the $g_{\sigma\tau}$ describe the gravitational field.

Thus, according to the general theory of relativity, gravitation occupies an exceptional position with regard to other forces, particularly the electromagnetic forces, since the ten functions representing the gravitational field at the same time define the metrical properties of the space measured.

B. Mathematical Aids to the Formulation of Generally Covariant Equations

Having seen in the foregoing that the general postulate of relativity leads to the requirement that the equations of physics shall be covariant in the face of any substitution of the co-ordinates $x_1 \ldots x_4$, we have to consider how such generally covariant equations can be found. We now turn to this purely mathematical task, and we shall find that in its solution a fundamental rôle is played by the invariant ds given in equation (3), which, borrowing from Gauss's theory of surfaces, we have called the "linear element."

The fundamental idea of this general theory of covariants is the following:—Let certain things ("tensors") be defined with respect to any system of co-ordinates by a number of functions of the co-ordinates, called the "components" of the tensor. There are then certain rules by which these components can be calculated for a new system of co-ordinates, if they are known for the original system of co-ordinates, and if the transformation connecting the two systems is known. The things hereafter called tensors are further characterized by the fact that the equations of transformation for their components are linear and homogeneous. Accordingly, all the components in the new system vanish, if they all vanish in the original system. If, therefore, a law of nature is expressed by equating all the components of a tensor to zero, it is generally covariant. By examining the laws of the formation of tensors, we acquire the means of formulating generally covariant laws.

§ 5. Contravariant and Covariant Four-vectors

Contravariant Four-vectors.—The linear element is defined by the four "components" dx_ν, for which the law of transformation is expressed by the equation

$$dx'_\sigma = \sum_\nu \frac{\partial x'_\sigma}{\partial x_\nu} dx_\nu \quad . \quad . \quad . \quad . \quad (5)$$

The dx'_σ are expressed as linear and homogeneous functions of the dx_ν. Hence we may look upon these co-ordinate differentials as the components of a "tensor" of the particular kind which we call a contravariant four-vector. Any thing which is defined relatively to the system of co-ordinates by four quantities A^ν, and which is transformed by the same law

$$A'^\sigma = \sum_\nu \frac{\partial x'_\sigma}{\partial x_\nu} A^\nu, \quad . \quad . \quad . \quad . \quad (5a)$$

we also call a contravariant four-vector. From (5a) it follows at once that the sums $A^\sigma \pm B^\sigma$ are also components of a four-vector, if A^σ and B^σ are such. Corresponding relations hold for all "tensors" subsequently to be introduced. (Rule for the addition and subtraction of tensors.)

Covariant Four-vectors.—We call four quantities A_ν the components of a covariant four-vector, if for any arbitrary choice of the contravariant four-vector B^ν

$$\sum_\nu A_\nu B^\nu = \text{Invariant} \quad . \quad . \quad . \quad (6)$$

The law of transformation of a covariant four-vector follows from this definition. For if we replace B^ν on the right-hand side of the equation

$$\sum_\sigma A'_\sigma B'^\sigma = \sum_\nu A_\nu B^\nu$$

by the expression resulting from the inversion of (5a),

$$\sum_\sigma \frac{\partial x_\nu}{\partial x'_\sigma} B'^\sigma,$$

we obtain

$$\sum_\sigma B'^\sigma \sum_\nu \frac{\partial x_\nu}{\partial x'_\sigma} A_\nu = \sum_\sigma B'^\sigma A'_\sigma .$$

Since this equation is true for arbitrary values of the B'^σ, it follows that the law of transformation is

$$A'_\sigma = \sum_\nu \frac{\partial x_\nu}{\partial x'_\sigma} A_\nu \quad . \quad . \quad . \quad . \quad (7)$$

Note on a Simplified Way of Writing the Expressions.—A glance at the equations of this paragraph shows that there is always a summation with respect to the indices which occur twice under a sign of summation (e.g. the index ν in (5)), and only with respect to indices which occur twice. It is therefore possible, without loss of clearness, to omit the sign of summation. In its place we introduce the convention:—If an index occurs twice in one term of an expression, it is always to be summed unless the contrary is expressly stated.

The difference between covariant and contravariant four-vectors lies in the law of transformation ((7) or (5) respectively). Both forms are tensors in the sense of the general remark above. Therein lies their importance. Following Ricci and

Levi-Civita, we denote the contravariant character by placing the index above, the covariant by placing it below.

§ 6. Tensors of the Second and Higher Ranks

Contravariant Tensors.—If we form all the sixteen products $A^{\mu\nu}$ of the components A^{μ} and B^{ν} of two contravariant four-vectors

$$A^{\mu\nu} = A^{\mu}B^{\nu} \quad . \quad . \quad . \quad . \quad (8)$$

then by (8) and (5a) $A^{\mu\nu}$ satisfies the law of transformation

$$A'^{\sigma\tau} = \frac{\partial x'_{\sigma}}{\partial x_{\mu}}\frac{\partial x'_{\tau}}{\partial x_{\nu}}A^{\mu\nu} \quad . \quad . \quad (9)$$

We call a thing which is described relatively to any system of reference by sixteen quantities, satisfying the law of transformation (9), a contravariant tensor of the second rank. Not every such tensor allows itself to be formed in accordance with (8) from two four-vectors, but it is easily shown that any given sixteen $A^{\mu\nu}$ can be represented as the sums of the $A^{\mu}B^{\nu}$ of four appropriately selected pairs of four-vectors. Hence we can prove nearly all the laws which apply to the tensor of the second rank defined by (9) in the simplest manner by demonstrating them for the special tensors of the type (8).

Contravariant Tensors of Any Rank.—It is clear that, on the lines of (8) and (9), contravariant tensors of the third and higher ranks may also be defined with 4^3 components, and so on. In the same way it follows from (8) and (9) that the contravariant four-vector may be taken in this sense as a contravariant tensor of the first rank.

Covariant Tensors.—On the other hand, if we take the sixteen products $A_{\mu\nu}$ of two covariant four-vectors A_{μ} and B_{ν},

$$A_{\mu\nu} = A_{\mu}B_{\nu}, \quad . \quad . \quad . \quad . \quad (10)$$

the law of transformation for these is

$$A'_{\sigma\tau} = \frac{\partial x_{\mu}}{\partial x'_{\sigma}}\frac{\partial x_{\nu}}{\partial x'_{\tau}}A_{\mu\nu} \quad . \quad . \quad . \quad (11)$$

This law of transformation defines the covariant tensor of the second rank. All our previous remarks on contravariant tensors apply equally to covariant tensors.

NOTE.—It is convenient to treat the scalar (or invariant) both as a contravariant and a covariant tensor of zero rank.

Mixed Tensors.—We may also define a tensor of the second rank of the type

$$A_{\mu}^{\nu} = A_{\mu}B^{\nu} \quad . \quad . \quad . \quad . \quad (12)$$

which is covariant with respect to the index μ, and contravariant with respect to the index ν. Its law of transformation is

$$A'^{\tau}_{\sigma} = \frac{\partial x'_{\tau}}{\partial x_{\nu}} \frac{\partial x_{\mu}}{\partial x'_{\sigma}} A_{\mu}^{\nu} \quad . \quad . \quad . \quad (13)$$

Naturally there are mixed tensors with any number of indices of covariant character, and any number of indices of contravariant character. Covariant and contravariant tensors may be looked upon as special cases of mixed tensors.

Symmetrical Tensors.—A contravariant, or a covariant tensor, of the second or higher rank is said to be symmetrical if two components, which are obtained the one from the other by the interchange of two indices, are equal. The tensor $A^{\mu\nu}$, or the tensor $A_{\mu\nu}$, is thus symmetrical if for any combination of the indices μ, ν,

$$A^{\mu\nu} = A^{\nu\mu}, \quad . \quad . \quad . \quad . \quad (14)$$

or respectively,

$$A_{\mu\nu} = A_{\nu\mu}. \quad . \quad . \quad . \quad . \quad (14a)$$

It has to be proved that the symmetry thus defined is a property which is independent of the system of reference. It follows in fact from (9), when (14) is taken into consideration, that

$$A'^{\sigma\tau} = \frac{\partial x'_{\sigma}}{\partial x_{\mu}} \frac{\partial x'_{\tau}}{\partial x_{\nu}} A^{\mu\nu} = \frac{\partial x'_{\sigma}}{\partial x_{\mu}} \frac{\partial x'_{\tau}}{\partial x_{\nu}} A^{\nu\mu} = \frac{\partial x'_{\sigma}}{\partial x_{\nu}} \frac{\partial x'_{\tau}}{\partial x_{\mu}} A^{\mu\nu} = A'^{\tau\sigma}.$$

The last equation but one depends upon the interchange of the summation indices μ and ν, i.e. merely on a change of notation.

Antisymmetrical Tensors.—A contravariant or a covariant tensor of the second, third, or fourth rank is said to be antisymmetrical if two components, which are obtained the one from the other by the interchange of two indices, are equal and of opposite sign. The tensor $A^{\mu\nu}$, or the tensor $A_{\mu\nu}$, is therefore antisymmetrical, if always

$$A^{\mu\nu} = -A^{\nu\mu}, \quad . \quad . \quad . \quad (15)$$

or respectively,

$$A_{\mu\nu} = -A_{\nu\mu} \quad . \quad . \quad . \quad (15a)$$

Of the sixteen components $A^{\mu\nu}$, the four components $A^{\mu\mu}$ vanish; the rest are equal and of opposite sign in pairs, so that there are only six components numerically different (a six-vector). Similarly we see that the antisymmetrical tensor of the third rank $A^{\mu\nu\sigma}$ has only four numerically different components, while the antisymmetrical tensor $A^{\mu\nu\sigma\tau}$ has only one. There are no antisymmetrical tensors of higher rank than the fourth in a continuum of four dimensions.

§ 7. Multiplication of Tensors

Outer Multiplication of Tensors.—We obtain from the components of a tensor of rank n and of a tensor of rank m the components of a tensor of rank $n + m$ by multiplying each component of the one tensor by each component of the other. Thus, for example, the tensors T arise out of the tensors A and B of different kinds,

$$T_{\mu\nu\sigma} = A_{\mu\nu}B_{\sigma},$$
$$T^{\mu\nu\sigma\tau} = A^{\mu\nu}B^{\sigma\tau},$$
$$T^{\sigma\tau}_{\mu\nu} = A_{\mu\nu}B^{\sigma\nu}.$$

The proof of the tensor character of T is given directly by the representations (8), (10), (12), or by the laws of transformation (9), (11), (13). The equations (8), (10), (12) are themselves examples of outer multiplication of tensors of the first rank.

"Contraction" of a Mixed Tensor.—From any mixed tensor we may form a tensor whose rank is less by two, by equating an index of covariant with one of contravariant character, and summing with respect to this index ("contraction"). Thus, for example, from the mixed tensor of the fourth rank $A^{\sigma\tau}_{\mu\nu}$, we obtain the mixed tensor of the second rank,

$$A^{\tau}_{\nu} = A^{\mu\tau}_{\mu\nu} \quad \left(= \sum_{\mu} A^{\mu\tau}_{\mu\nu} \right),$$

and from this, by a second contraction, the tensor of zero rank,

$$A = A^{\nu}_{\nu} = A^{\mu\nu}_{\mu\nu}.$$

The proof that the result of contraction really possesses the tensor character is given either by the representation of a tensor according to the generalization of (12) in combination with (6), or by the generalization of (13).

Inner and Mixed Multiplication of Tensors.—These consist in a combination of outer multiplication with contraction.

Examples.—From the covariant tensor of the second rank $A_{\mu\nu}$ and the contravariant tensor of the first rank B^{σ} we form by outer multiplication the mixed tensor

$$D^{\sigma}_{\mu\nu} = A_{\mu\nu}B^{\sigma}.$$

On contraction with respect to the indices ν and σ, we obtain the covariant four-vector

$$D_{\mu} = D^{\nu}_{\mu\nu} = A_{\mu\nu}B^{\nu}.$$

This we call the inner product of the tensors $A_{\mu\nu}$ and B^{σ}. Analogously we form from the tensors $A_{\mu\nu}$ and $B^{\sigma\tau}$, by outer multiplication and double contraction, the inner product $A_{\mu\nu}B^{\mu\nu}$. By outer multiplication and one contraction, we obtain from $A_{\mu\nu}$ and $B^{\sigma\tau}$ the mixed tensor of the second rank $D^{\tau}_{\mu} = A_{\mu\nu}B^{\nu\tau}$. This operation may be aptly characterized as a mixed one, being "outer" with respect to the indices μ and τ, and "inner" with respect to the indices ν and σ.

We now prove a proposition which is often useful as evidence of tensor character. From what has just been explained, $A_{\mu\nu}B^{\mu\nu}$ is a scalar if $A_{\mu\nu}$ and $B^{\sigma\tau}$ are tensors. But we may also make the following assertion: If $A_{\mu\nu}B^{\mu\nu}$ is a scalar *for any choice of the tensor* $B^{\mu\nu}$, then $A_{\mu\nu}$ has tensor character. For, by hypothesis, for any substitution,

$$A'_{\sigma\tau}B'^{\sigma\tau} = A_{\mu\nu}B^{\mu\nu}.$$

But by an inversion of (9)

$$B^{\mu\nu} = \frac{\partial x_{\mu}}{\partial x'_{\sigma}}\frac{\partial x_{\nu}}{\partial x'_{\tau}}B'^{\sigma\tau}.$$

This, inserted in the above equation, gives

$$\left(A'_{\sigma\tau} - \frac{\partial x_{\mu}}{\partial x'_{\sigma}}\frac{\partial x_{\nu}}{\partial x'_{\tau}}A_{\mu\nu}\right)B'^{\sigma\tau} = 0.$$

This can only be satisfied for arbitrary values of $B'^{\sigma\tau}$ if the bracket vanishes. The result then follows by equation (11). This rule applies correspondingly to tensors of any rank and character, and the proof is analogous in all cases.

The rule may also be demonstrated in this form: If B^μ and C^ν are any vectors, and if, for all values of these, the inner product $A_{\mu\nu}B^\mu C^\nu$ is a scalar, then $A_{\mu\nu}$ is a covariant tensor. This latter proposition also holds good even if only the more special assertion is correct, that with any choice of the four-vector B^μ the inner product $A_{\mu\nu}B^\mu B^\nu$ is a scalar, if in addition it is known that $A_{\mu\nu}$ satisfies the condition of symmetry $A_{\mu\nu} = A_{\nu\mu}$. For by the method given above we prove the tensor character of $(A_{\mu\nu} + A_{\nu\mu})$, and from this the tensor character of $A_{\mu\nu}$ follows on account of symmetry. This also can be easily generalized to the case of covariant and contravariant tensors of any rank.

Finally, there follows from what has been proved, this law, which may also be generalized for any tensors: If for any choice of the four-vector B^ν the quantities $A_{\mu\nu}B^\nu$ form a tensor of the first rank, then $A_{\mu\nu}$ is a tensor of the second rank. For, if C^μ is any four-vector, then on account of the tensor character of $A_{\mu\nu}B^\nu$, the inner product $A_{\mu\nu}B^\nu C^\mu$ is a scalar for any choice of the two four-vectors B^ν and C^μ. From which the proposition follows.

§ 8. Some Aspects of the Fundamental Tensor $g_{\mu\nu}$

The Covariant Fundamental Tensor.—In the invariant expression for the square of the linear element,

$$ds^2 = g_{\mu\nu}dx_\mu dx_\nu,$$

the part played by the dx_μ is that of a contravariant vector which may be chosen at will. Since further, $g_{\mu\nu} = g_{\nu\mu}$, it follows from the considerations of the preceding paragraph that $g_{\mu\nu}$ is a covariant tensor of the second rank. We call it the "fundamental tensor." In what follows we deduce some properties of this tensor which, it is true, apply to any tensor of the second rank. But as the fundamental tensor plays a special part in our theory, which has its physical basis in the peculiar effects of gravitation, it so happens that the relations to be developed are of importance to us only in the case of the fundamental tensor.

The Contravariant Fundamental Tensor.—If in the determinant formed by the elements $g_{\mu\nu}$, we take the co-factor of each of the $g_{\mu\nu}$ and divide it by the determinant $g = | g_{\mu\nu} |$, we obtain certain quantities $g^{\mu\nu} (= g^{\nu\mu})$ which, as we shall demonstrate, form a contravariant tensor.

By a known property of determinants

$$g_{\mu\sigma}g^{\nu\sigma} = \delta_{\mu}^{\nu} \quad . \quad . \quad . \quad . \quad (16)$$

where the symbol δ_{μ}^{ν} denotes 1 or 0, according as $\mu = \nu$ or $\mu \neq \nu$.

Instead of the above expression for ds^2 we may thus write

$$g_{\mu\sigma}\delta_{\nu}^{\sigma}dx_{\mu}dx_{\nu}$$

or, by (16)

$$g_{\mu\sigma}g_{\nu\tau}g^{\sigma\tau}dx_{\mu}dx_{\nu}.$$

But, by the multiplication rules of the preceding paragraphs, the quantities

$$d\xi_{\sigma} = g_{\mu\sigma}dx_{\mu}$$

form a covariant four-vector, and in fact an arbitrary vector, since the dx_{μ} are arbitrary. By introducing this into our expression we obtain

$$ds^2 = g^{\sigma\tau}d\xi_{\sigma}d\xi_{\tau}.$$

Since this, with the arbitrary choice of the vector $d\xi_{\sigma}$, is a scalar, and $g^{\sigma\tau}$ by its definition is symmetrical in the indices σ and τ, it follows from the results of the preceding paragraph that $g^{\sigma\tau}$ is a contravariant tensor.

It further follows from (16) that δ_{μ} is also a tensor, which we may call the mixed fundamental tensor.

The Determinant of the Fundamental Tensor.—By the rule for the multiplication of determinants

$$| g_{\mu\alpha}g^{\alpha\nu} | = | g_{\mu\alpha} | \times | g^{\alpha\nu} |.$$

On the other hand

$$| g_{\mu\alpha}g^{\alpha\nu} | = | \delta_{\mu}^{\nu} | = 1.$$

It therefore follows that

$$| g_{\mu\nu} | \times | g^{\mu\nu} | = 1 \quad . \quad . \quad . \quad (17)$$

The Volume Scalar.—We seek first the law of transformation of the determinant $g = | g_{\mu\nu} |$. In accordance with (11)

$$g' = \left| \frac{\partial x_\mu}{\partial x'_\sigma} \frac{\partial x}{\partial x'_\tau} g_{\mu\nu} \right| .$$

Hence, by a double application of the rule for the multiplication of determinants, it follows that

$$g' = \left| \frac{\partial x_\mu}{\partial x'_\sigma} \right| \cdot \left| \frac{\partial x_\nu}{\partial x'_\tau} \right| \cdot | g_{\mu\nu} | = \left| \frac{\partial x_\mu}{\partial x'_\sigma} \right|^2 g,$$

or

$$\sqrt{g'} = \left| \frac{\partial x_\mu}{\partial x'_\sigma} \right| \sqrt{g}.$$

On the other hand, the law of transformation of the element of volume

$$d\tau = \int dx_1 dx_2 dx_3 dx_4$$

is, in accordance with the theorem of Jacobi,

$$d\tau' = \left| \frac{\partial x'_\sigma}{\partial x_\mu} \right| d\tau.$$

By multiplication of the last two equations, we obtain

$$\sqrt{g'} d\tau' = \sqrt{g} d\tau \quad . \quad . \quad . \quad (18).$$

Instead of $\sqrt{g}$, we introduce in what follows the quantity $\sqrt{-g}$, which is always real on account of the hyperbolic character of the space-time continuum. The invariant $\sqrt{-g}d\tau$ is equal to the magnitude of the four-dimensional element of volume in the "local" system of reference, as measured with rigid rods and clocks in the sense of the special theory of relativity.

Note on the Character of the Space-time Continuum.—Our assumption that the special theory of relativity can always be applied to an infinitely small region, implies that ds^2 can always be expressed in accordance with (1) by means of real quantities $dX_1 \ldots dX_4$. If we denote by $d\tau_0$ the "natural" element of volume dX_1, dX_2, dX_3, dX_4, then

$$d\tau_0 = \sqrt{-g}d\tau \quad . \quad . \quad . \quad (18a)$$

If $\sqrt{-g}$ were to vanish at a point of the four-dimensional continuum, it would mean that at this point an infinitely small "natural" volume would correspond to a finite volume in the co-ordinates. Let us assume that this is never the case. Then g cannot change sign. We will assume that, in the sense of the special theory of relativity, g always has a finite negative value. This is a hypothesis as to the physical nature of the continuum under consideration, and at the same time a convention as to the choice of co-ordinates.

But if $-g$ is always finite and positive, it is natural to settle the choice of co-ordinates *a posteriori* in such a way that this quantity is always equal to unity. We shall see later that by such a restriction of the choice of co-ordinates it is possible to achieve an important simplification of the laws of nature.

In place of (18), we then have simply $d\tau' = d\tau$, from which, in view of Jacobi's theorem, it follows that

$$\left| \frac{\partial x'_\sigma}{\partial x_\mu} \right| = 1 \quad . \quad . \quad . \quad . \quad (19)$$

Thus, with this choice of co-ordinates, only substitutions for which the determinant is unity are permissible.

But it would be erroneous to believe that this step indicates a partial abandonment of the general postulate of relativity. We do not ask "What are the laws of nature which are co-variant in face of all substitutions for which the determinant is unity?" but our question is "What are the generally co-variant laws of nature?" It is not until we have formulated these that we simplify their expression by a particular choice of the system of reference.

The Formation of New Tensors by Means of the Fundamental Tensor.—Inner, outer, and mixed multiplication of a tensor by the fundamental tensor give tensors of different character and rank. For example,

$$A^\mu = g^{\mu\sigma} A_\sigma,$$
$$A = g_{\mu\nu} A^{\mu\nu}.$$

The following forms may be specially noted:—

$$A^{\mu\nu} = g^{\mu\alpha} g^{\nu\beta} A_{\alpha\beta},$$
$$A_{\mu\nu} = g_{\mu\alpha} g_{\nu\beta} A^{\alpha\beta}$$

(the " complements " of covariant and contravariant tensors respectively), and

$$B_{\mu\nu} = g_{\mu\nu} g^{\alpha\beta} A_{\alpha\beta}.$$

We call $B_{\mu\nu}$ the reduced tensor associated with $A_{\mu\nu}$. Similarly,

$$B^{\mu\nu} = g^{\mu\nu} g_{\alpha\beta} A^{\alpha\beta}.$$

It may be noted that $g^{\mu\nu}$ is nothing more than the complement of $g_{\mu\nu}$, since

$$g^{\mu\alpha} g^{\nu\beta} g_{\alpha\beta} = g^{\mu\alpha} \delta_\alpha^\nu = g^{\mu\nu}.$$

§ 9. The Equation of the Geodetic Line. The Motion of a Particle

As the linear element ds is defined independently of the system of co-ordinates, the line drawn between two points P and P′ of the four-dimensional continuum in such a way that $\int ds$ is stationary—a geodetic line—has a meaning which also is independent of the choice of co-ordinates. Its equation is

$$\delta \int_P^{P'} ds = 0 \quad . \quad . \quad . \quad (20)$$

Carrying out the variation in the usual way, we obtain from this equation four differential equations which define the geodetic line; this operation will be inserted here for the sake of completeness. Let λ be a function of the co-ordinates x_ν, and let this define a family of surfaces which intersect the required geodetic line as well as all the lines in immediate proximity to it which are drawn through the points P and P′. Any such line may then be supposed to be given by expressing its co-ordinates x_ν as functions of λ. Let the symbol δ indicate the transition from a point of the required geodetic to the point corresponding to the same λ on a neighbouring line. Then for (20) we may substitute

$$\left.\begin{aligned} \int_{\lambda_1}^{\lambda_2} \delta w \, d\lambda &= 0 \\ w^2 &= g_{\mu\nu} \frac{dx_\mu}{d\lambda} \frac{dx_\nu}{d\lambda} \end{aligned}\right\} \quad . \quad . \quad . \quad (20a)$$

But since

$$\delta w = \frac{1}{w}\left\{\frac{1}{2}\frac{\partial g_{\mu\nu}}{\partial x_\sigma}\frac{dx_\mu}{d\lambda}\frac{dx_\nu}{d\lambda}\delta x_\sigma + g_{\mu\nu}\frac{dx_\mu}{d\lambda}\delta\left(\frac{dx_\nu}{d\lambda}\right)\right\},$$

and

$$\delta\left(\frac{dx_\nu}{d\lambda}\right) = \frac{d}{d\lambda}(\delta x_\nu),$$

we obtain from (20a), after a partial integration,

$$\int_{\lambda_1}^{\lambda_2} \kappa_\sigma \delta x_\sigma d\lambda = 0,$$

where

$$\kappa_\sigma = \frac{d}{d\lambda}\left\{\frac{g_{\mu\nu}}{w}\frac{dx_\mu}{d\lambda}\right\} - \frac{1}{2w}\frac{\partial g_{\mu\nu}}{\partial x_\sigma}\frac{dx_\mu}{d\lambda}\frac{dx_\nu}{d\lambda} \quad . \quad (20b)$$

Since the values of δx_σ are arbitrary, it follows from this that

$$\kappa_\sigma = 0 \quad . \quad . \quad . \quad . \quad (20c)$$

are the equations of the geodetic line.

If ds does not vanish along the geodetic line we may choose the "length of the arc" s, measured along the geodetic line, for the parameter λ. Then $w = 1$, and in place of (20c) we obtain

$$g_{\mu\nu}\frac{d^2 x_\mu}{ds^2} + \frac{\partial g_{\mu\nu}}{\partial x_\sigma}\frac{dx_\sigma}{ds}\frac{dx_\mu}{ds} - \frac{1}{2}\frac{\partial g_{\mu\nu}}{\partial x_\sigma}\frac{dx_\mu}{ds}\frac{dx_\nu}{ds} = 0$$

or, by a mere change of notation,

$$g_{\alpha\sigma}\frac{d^2 x_\alpha}{ds^2} + [\mu\nu, \sigma]\frac{dx_\mu}{ds}\frac{dx_\nu}{ds} = 0 \quad . \quad . \quad (20d)$$

where, following Christoffel, we have written

$$[\mu\nu, \sigma] = \frac{1}{2}\left(\frac{\partial g_{\mu\sigma}}{\partial x_\nu} + \frac{\partial g_{\nu\sigma}}{\partial x_\mu} - \frac{\partial g_{\mu\nu}}{\partial x_\sigma}\right) \quad . \quad (21)$$

Finally, if we multiply (20d) by $g^{\sigma\tau}$ (outer multiplication with respect to τ, inner with respect to σ), we obtain the equations of the geodetic line in the form

$$\frac{d^2 x_\tau}{ds^2} + \{\mu\nu, \tau\}\frac{dx_\mu}{ds}\frac{dx_\nu}{ds} = 0 \quad . \quad . \quad . \quad (22)$$

where, following Christoffel, we have set

$$\{\mu\nu, \tau\} = g^{\tau\alpha}[\mu\nu, \alpha] \quad . \quad . \quad . \quad (23)$$

§ 10. The Formation of Tensors by Differentiation

With the help of the equation of the geodetic line we can now easily deduce the laws by which new tensors can be formed from old by differentiation. By this means we are able for the first time to formulate generally covariant differential equations. We reach this goal by repeated application of the following simple law :—

If in our continuum a curve is given, the points of which are specified by the arcual distance s measured from a fixed point on the curve, and if, further, ϕ is an invariant function of space, then $d\phi/ds$ is also an invariant. The proof lies in this, that ds is an invariant as well as $d\phi$.

As

$$\frac{d\phi}{ds} = \frac{\partial\phi}{\partial x_\mu}\frac{dx_\mu}{ds}$$

therefore

$$\psi = \frac{\partial\phi}{dx_\mu}\frac{dx_\mu}{ds}$$

is also an invariant, and an invariant for all curves starting from a point of the continuum, that is, for any choice of the vector dx_μ. Hence it immediately follows that

$$A_\mu = \frac{\partial\phi}{\partial x_\mu} \quad . \quad . \quad . \quad . \quad (24)$$

is a covariant four-vector—the " gradient " of ϕ.

According to our rule, the differential quotient

$$\chi = \frac{d\psi}{ds}$$

taken on a curve, is similarly an invariant. Inserting the value of ψ, we obtain in the first place

$$\chi = \frac{\partial^2\phi}{\partial x_\mu \partial x_\nu}\frac{dx_\mu}{ds}\frac{dx_\nu}{ds} + \frac{\partial\phi}{\partial x_\mu}\frac{d^2x_\mu}{ds^2}.$$

The existence of a tensor cannot be deduced from this forthwith. But if we may take the curve along which we have differentiated to be a geodetic, we obtain on substitution for d^2x_ν/ds^2 from (22),

$$\chi = \left(\frac{\partial^2\phi}{\partial x_\mu \partial x_\nu} - \{\mu\nu, \tau\}\frac{\partial\phi}{\partial x_\tau}\right)\frac{dx_\mu}{ds}\frac{dx_\nu}{ds}.$$

Since we may interchange the order of the differentiations,

and since by (23) and (21) $\{\mu\nu, \tau\}$ is symmetrical in μ and ν, it follows that the expression in brackets is symmetrical in μ and ν. Since a geodetic line can be drawn in any direction from a point of the continuum, and therefore dx_μ/ds is a four-vector with the ratio of its components arbitrary, it follows from the results of § 7 that

$$A_{\mu\nu} = \frac{\partial^2\phi}{\partial x_\mu \partial x_\nu} - \{\mu\nu, \tau\}\frac{\partial\phi}{\partial x_\tau} \quad . \quad . \quad . \quad (25)$$

is a covariant tensor of the second rank. We have therefore come to this result: from the covariant tensor of the first rank

$$A_\mu = \frac{\partial\phi}{\partial x_\mu}$$

we can, by differentiation, form a covariant tensor of the second rank

$$A_{\mu\nu} = \frac{\partial A_\mu}{\partial x_\nu} - \{\mu\nu, \tau\}A_\tau \quad . \quad . \quad . \quad (26)$$

We call the tensor $A_{\mu\nu}$ the " extension " (covariant derivative) of the tensor A_μ. In the first place we can readily show that the operation leads to a tensor, even if the vector A_μ cannot be represented as a gradient. To see this, we first observe that

$$\psi\frac{\partial\phi}{\partial x_\mu}$$

is a covariant vector, if ψ and ϕ are scalars. The sum of four such terms

$$S_\mu = \psi^{(1)}\frac{\phi\partial^{(1)}}{\partial x_\mu} + . + . + \psi^{(4)}\frac{\partial\phi^{(4)}}{\partial x_\mu},$$

is also a covariant vector, if $\psi^{(1)}, \phi^{(1)} \ . \ . \ . \ \psi^{(4)}, \phi^{(4)}$ are scalars. But it is clear that any covariant vector can be represented in the form S_μ. For, if A_μ is a vector whose components are any given functions of the x_ν, we have only to put (in terms of the selected system of co-ordinates)

$$\begin{aligned} \psi^{(1)} &= A_1, \quad \phi^{(1)} = x_1, \\ \psi^{(2)} &= A_2, \quad \phi^{(2)} = x_2, \\ \psi^{(3)} &= A_3, \quad \phi^{(3)} = x_3, \\ \psi^{(4)} &= A_4, \quad \phi^{(4)} = x_4, \end{aligned}$$

in order to ensure that S_μ shall be equal to A_μ.

Therefore, in order to demonstrate that $A_{\mu\nu}$ is a tensor if *any* covariant vector is inserted on the right-hand side for A_μ, we only need show that this is so for the vector S_μ. But for this latter purpose it is sufficient, as a glance at the right-hand side of (26) teaches us, to furnish the proof for the case

$$A_\mu = \psi \frac{\partial \phi}{\partial x_\mu}.$$

Now the right-hand side of (25) multiplied by ψ,

$$\psi \frac{\partial^2 \phi}{\partial x_\mu \partial x_\nu} - \{\mu\nu, \tau\}\psi \frac{\partial \phi}{\partial x_\tau}$$

is a tensor. Similarly

$$\frac{\partial \psi}{\partial x_\mu} \frac{\partial \phi}{\partial x_\nu}$$

being the outer product of two vectors, is a tensor. By addition, there follows the tensor character of

$$\frac{\partial}{\partial x_\nu}\left(\psi \frac{\partial \phi}{\partial x_\mu}\right) - \{\mu\nu, \tau\}\left(\psi \frac{\partial \phi}{\partial x_\tau}\right).$$

As a glance at (26) will show, this completes the demonstration for the vector

$$\psi \frac{\partial \phi}{\partial x_\mu}$$

and consequently, from what has already been proved, for any vector A_μ.

By means of the extension of the vector, we may easily define the "extension" of a covariant tensor of any rank. This operation is a generalization of the extension of a vector. We restrict ourselves to the case of a tensor of the second rank, since this suffices to give a clear idea of the law of formation.

As has already been observed, any covariant tensor of the second rank can be represented * as the sum of tensors of the

* By outer multiplication of the vector with arbitrary components A_{11}, A_{12}, A_{13}, A_{14} by the vector with components 1, 0, 0, 0, we produce a tensor with components

$$\begin{matrix} A_{11} & A_{12} & A_{13} & A_{14} \\ 0 & 0 & 0 & 0 \\ 0 & 0 & 0 & 0 \\ 0 & 0 & 0 & 0. \end{matrix}$$

By the addition of four tensors of this type, we obtain the tensor $A_{\mu\nu}$ with any ssigned components.

type $A_\mu B_\nu$. It will therefore be sufficient to deduce the expression for the extension of a tensor of this special type. By (26) the expressions

$$\frac{\partial A_\mu}{\partial x_\sigma} - \{\sigma\mu, \tau\}A_\tau,$$

$$\frac{\partial B_\nu}{\partial x_\sigma} - \{\sigma\nu, \tau\}B_\tau,$$

are tensors. On outer multiplication of the first by B_ν, and of the second by A_μ, we obtain in each case a tensor of the third rank. By adding these, we have the tensor of the third rank

$$A_{\mu\nu\sigma} = \frac{\partial A_{\mu\nu}}{\partial x_\sigma} - \{\sigma\mu, \tau\}A_{\tau\nu} - \{\sigma\nu, \tau\}A_{\mu\tau} \quad . \quad . \quad (27)$$

where we have put $A_{\mu\nu} = A_\mu B_\nu$. As the right-hand side of (27) is linear and homogeneous in the $A_{\mu\nu}$ and their first derivatives, this law of formation leads to a tensor, not only in the case of a tensor of the type $A_\mu B_\nu$, but also in the case of a sum of such tensors, i.e. in the case of any covariant tensor of the second rank. We call $A_{\mu\nu\sigma}$ the extension of the tensor $A_{\mu\nu}$.

It is clear that (26) and (24) concern only special cases of extension (the extension of the tensors of rank one and zero respectively).

In general, all special laws of formation of tensors are included in (27) in combination with the multiplication of tensors.

§ 11. Some Cases of Special Importance

The Fundamental Tensor.—We will first prove some lemmas which will be useful hereafter. By the rule for the differentiation of determinants

$$dg = g^{\mu\nu} g dg_{\mu\nu} = - g_{\mu\nu} g dg^{\mu\nu} \quad . \quad . \quad (28)$$

The last member is obtained from the last but one, if we bear in mind that $g_{\mu\nu}g^{\mu'\nu} = \delta^{\mu'}_{\mu}$, so that $g_{\mu\nu}g^{\mu\nu} = 4$, and consequently

$$g_{\mu\nu}dg^{\mu\nu} + g^{\mu\nu}dg_{\mu\nu} = 0.$$

From (28), it follows that

$$\frac{1}{\sqrt{-g}}\frac{\partial\sqrt{-g}}{\partial x_\sigma} = \frac{1}{2}\frac{\partial \log(-g)}{\partial x_\sigma} = \frac{1}{2}g^{\mu\nu}\frac{\partial g_{\mu\nu}}{\partial x_\sigma} = \frac{1}{2}g_{\mu\nu}\frac{\partial g^{\mu\nu}}{\partial x_\sigma}. \quad (29)$$

Further, from $g_{\mu\sigma}g^{\nu\sigma} = \delta^\nu_\mu$, it follows on differentiation that

$$\left.\begin{aligned} g_{\mu\sigma}dg^{\nu\sigma} &= -g^{\nu\sigma}dg_{\mu\sigma} \\ g_{\mu\sigma}\frac{\partial g^{\nu\sigma}}{\partial x_\lambda} &= -g^{\nu\sigma}\frac{\partial g_{\mu\sigma}}{\partial x_\lambda} \end{aligned}\right\} \quad . \quad . \quad (30)$$

From these, by mixed multiplication by $g^{\sigma\tau}$ and $g_{\nu\lambda}$ respectively, and a change of notation for the indices, we have

$$\left.\begin{aligned} dg^{\mu\nu} &= -g^{\mu\alpha}g^{\nu\beta}dg_{\alpha\beta} \\ \frac{\partial g^{\mu\nu}}{\partial x_\sigma} &= -g^{\mu\alpha}g^{\nu\beta}\frac{\partial g_{\alpha\beta}}{\partial x_\sigma} \end{aligned}\right\} \quad . \quad . \quad (31)$$

and

$$\left.\begin{aligned} dg_{\mu\nu} &= -g_{\mu\alpha}g_{\nu\beta}dg^{\alpha\beta} \\ \frac{\partial g_{\mu\nu}}{\partial x_\sigma} &= -g_{\mu\alpha}g_{\nu\beta}\frac{\partial g^{\alpha\beta}}{\partial x_\sigma} \end{aligned}\right\} \quad . \quad . \quad . \quad (32)$$

The relation (31) admits of a transformation, of which we also have frequently to make use. From (21)

$$\frac{\partial g_{\alpha\beta}}{\partial x_\sigma} = [\alpha\sigma, \beta] + [\beta\sigma, \alpha] \quad . \quad . \quad . \quad (33)$$

Inserting this in the second formula of (31), we obtain, in view of (23)

$$\frac{\partial g^{\mu\nu}}{\partial x_\sigma} = -g^{\mu\tau}\{\tau\sigma, \nu\} - g^{\nu\tau}\{\tau\sigma, \mu\} \quad . \quad . \quad (34)$$

Substituting the right-hand side of (34) in (29), we have

$$\frac{1}{\sqrt{-g}}\frac{\partial\sqrt{-g}}{\partial x_\sigma} = \{\mu\sigma, \mu\} \quad . \quad . \quad (29a)$$

The "Divergence" of a Contravariant Vector.—If we take the inner product of (26) by the contravariant fundamental tensor $g^{\mu\nu}$, the right-hand side, after a transformation of the first term, assumes the form

$$\frac{\partial}{\partial x_\nu}(g^{\mu\nu}A_\mu) - A_\mu\frac{\partial g^{\mu\nu}}{\partial x_\nu} - \frac{1}{2}g^{\tau\alpha}\left(\frac{\partial g_{\mu\alpha}}{\partial x_\nu} + \frac{\partial g_{\nu\alpha}}{\partial x_\mu} - \frac{\partial g_{\mu\nu}}{\partial x_\alpha}\right)g^{\mu\nu}A_\tau.$$

In accordance with (31) and (29), the last term of this expression may be written

$$\frac{1}{2}\frac{\partial g^{\tau\nu}}{\partial x_\nu}A_\tau + \frac{1}{2}\frac{\partial g^{\tau\mu}}{\partial x_\mu}A_\tau + \frac{1}{\sqrt{-g}}\frac{\partial\sqrt{-g}}{\partial x_\alpha}g^{\mu\nu}A_\tau.$$

As the symbols of the indices of summation are immaterial, the first two terms of this expression cancel the second of the one above. If we then write $g^{\mu\nu}A_\mu = A^\nu$, so that A^ν like A_μ is an arbitrary vector, we finally obtain

$$\Phi = \frac{1}{\sqrt{-g}}\frac{\partial}{\partial x_\nu}(\sqrt{-g}A^\nu) . \quad . \quad . \quad (35)$$

This scalar is the *divergence* of the contravariant vector A^ν.

The "Curl" of a Covariant Vector.—The second term in (26) is symmetrical in the indices μ and ν. Therefore $A_{\mu\nu} - A_{\nu\mu}$ is a particularly simply constructed antisymmetrical tensor. We obtain

$$B_{\mu\nu} = \frac{\partial A_\mu}{\partial x_\nu} - \frac{\partial A_\nu}{\partial x_\mu} \quad . \quad . \quad . \quad (36)$$

Antisymmetrical Extension of a Six-vector.—Applying (27) to an antisymmetrical tensor of the second rank $A_{\mu\nu}$, forming in addition the two equations which arise through cyclic permutations of the indices, and adding these three equations, we obtain the tensor of the third rank

$$B_{\mu\nu\sigma} = A_{\mu\nu\sigma} + A_{\nu\sigma\mu} + A_{\sigma\mu\nu} = \frac{\partial A_{\mu\nu}}{\partial x_\sigma} + \frac{\partial A_{\nu\sigma}}{\partial x_\mu} + \frac{\partial A_{\sigma\mu}}{\partial x_\nu} \quad (37)$$

which it is easy to prove is antisymmetrical.

The Divergence of a Six-vector.—Taking the mixed product of (27) by $g^{\mu\alpha}g^{\nu\beta}$, we also obtain a tensor. The first term on the right-hand side of (27) may be written in the form

$$\frac{\partial}{\partial x_\sigma}(g^{\mu\alpha}g^{\nu\beta}A_{\mu\nu}) - g^{\mu\alpha}\frac{\partial g^{\nu\beta}}{\partial x_\sigma}A_{\mu\nu} - g^{\nu\beta}\frac{\partial g^{\mu\alpha}}{\partial x_\sigma}A_{\mu\nu}.$$

If we write $A^{\alpha\beta}_\sigma$ for $g^{\mu\alpha}g^{\nu\beta}A_{\mu\nu\sigma}$ and $A^{\alpha\beta}$ for $g^{\mu\alpha}g^{\nu\beta}A_{\mu\nu}$, and in the transformed first term replace

$$\frac{\partial g^{\nu\beta}}{\partial x_\sigma} \text{ and } \frac{\partial g^{\mu\alpha}}{\partial x_\sigma}$$

by their values as given by (34), there results from the right-hand side of (27) an expression consisting of seven terms, of which four cancel, and there remains

$$A^{\alpha\beta}_{\sigma} = \frac{\partial A^{\alpha\beta}}{\partial x_\sigma} + \{\sigma\gamma, \alpha\}A^{\gamma\beta} + \{\sigma\gamma, \beta\}A^{\alpha\gamma} . \quad . \quad (38)$$

This is the expression for the extension of a contravariant tensor of the second rank, and corresponding expressions for the extension of contravariant tensors of higher and lower rank may also be formed.

We note that in an analogous way we may also form the extension of a mixed tensor :—

$$A^{\alpha}_{\mu\sigma} = \frac{\partial A^{\alpha}_{\mu}}{\partial x_\sigma} - \{\sigma\mu, \tau\}A^{\alpha}_{\tau} + \{\sigma\tau, \alpha\}A^{\tau}_{\mu} . \quad . \quad (39)$$

On contracting (38) with respect to the indices β and σ (inner multiplication by δ^{σ}_{β}), we obtain the vector

$$A^{\alpha} = \frac{\partial A^{\alpha\beta}}{\partial x_\beta} + \{\beta\gamma, \beta\}A^{\alpha\gamma} + \{\beta\gamma, \alpha\}A^{\gamma\beta}.$$

On account of the symmetry of $\{\beta\gamma, \alpha\}$ with respect to the indices β and γ, the third term on the right-hand side vanishes, if $A^{\alpha\beta}$ is, as we will assume, an antisymmetrical tensor. The second term allows itself to be transformed in accordance with (29a). Thus we obtain

$$A^{\alpha} = \frac{1}{\sqrt{-g}} \frac{\partial(\sqrt{-g}A^{\alpha\beta})}{\partial x_\beta} . \quad . \quad . \quad (40)$$

This is the expression for the divergence of a contravariant six-vector.

The Divergence of a Mixed Tensor of the Second Rank.— Contracting (39) with respect to the indices α and σ, and taking (29a) into consideration, we obtain

$$\sqrt{-g}A_{\mu} = \frac{\partial(\sqrt{-g}A^{\sigma}_{\mu})}{\partial x_\sigma} - \{\sigma\mu, \tau\}\sqrt{-g}A^{\sigma}_{\tau} \quad . \quad (41)$$

If we introduce the contravariant tensor $A^{\rho\sigma} = g^{\rho\tau}A^{\sigma}_{\tau}$ in the last term, it assumes the form

$$-[\sigma\mu, \rho]\sqrt{-g}A^{\rho\sigma}.$$

If, further, the tensor $A^{\rho\sigma}$ is symmetrical, this reduces to

$$-\tfrac{1}{2}\sqrt{-g}\frac{\partial g_{\rho\sigma}}{\partial x_\mu}A^{\rho\sigma}.$$

Had we introduced, instead of $A^{\rho\sigma}$, the covariant tensor $A_{\rho\sigma} = g_{\rho\alpha}g_{\sigma\beta}A^{\alpha\beta}$, which is also symmetrical, the last term, by virtue of (31), would assume the form

$$\tfrac{1}{2}\sqrt{-g}\frac{\partial g^{\rho\sigma}}{\partial x_\mu}A_{\rho\sigma}.$$

In the case of symmetry in question, (41) may therefore be replaced by the two forms

$$\sqrt{-g}A_\mu = \frac{\partial(\sqrt{-g}A_\mu^\sigma)}{\partial x_\sigma} - \tfrac{1}{2}\frac{\partial g_{\rho\sigma}}{\partial x_\mu}\sqrt{-g}A^{\rho\sigma} \quad . \quad (41a)$$

$$\sqrt{-g}A_\mu = \frac{\partial(\sqrt{-g}A_\mu^\sigma)}{\partial x_\sigma} + \tfrac{1}{2}\frac{\partial g^{\rho\sigma}}{\partial x_\mu}\sqrt{-g}A_{\rho\sigma} \quad . \quad (41b)$$

which we have to employ later on.

§ 12. The Riemann-Christoffel Tensor

We now seek the tensor which can be obtained from the fundamental tensor *alone*, by differentiation. At first sight the solution seems obvious. We place the fundamental tensor of the $g_{\mu\nu}$ in (27) instead of any given tensor $A_{\mu\nu}$, and thus have a new tensor, namely, the extension of the fundamental tensor. But we easily convince ourselves that this extension vanishes identically. We reach our goal, however, in the following way. In (27) place

$$A_{\mu\nu} = \frac{\partial A_\mu}{\partial x_\nu} - \{\mu\nu, \rho\}A_\rho,$$

i.e. the extension of the four-vector A_μ. Then (with a somewhat different naming of the indices) we get the tensor of the third rank

$$A_{\mu\sigma\tau} = \frac{\partial^2 A_\mu}{\partial x_\sigma \partial x_\tau} - \{\mu\sigma, \rho\}\frac{\partial A_\rho}{\partial x_\tau} - \{\mu\tau, \rho\}\frac{\partial A_\rho}{\partial x_\sigma} - \{\sigma\tau, \rho\}\frac{\partial A_\mu}{\partial x_\rho}$$

$$+ \left[-\frac{\partial}{\partial x_\tau}\{\mu\sigma, \rho\} + \{\mu\tau, \alpha\}\{\alpha\sigma, \rho\} + \{\sigma\tau, \alpha\}\{\alpha\mu, \rho\}\right]A_\rho.$$

This expression suggests forming the tensor $A_{\mu\sigma\tau} - A_{\mu\tau\sigma}$. For, if we do so, the following terms of the expression for $A_{\mu\sigma\tau}$ cancel those of $A_{\mu\tau\sigma}$, the first, the fourth, and the member corresponding to the last term in square brackets; because all these are symmetrical in σ and τ. The same holds good for the sum of the second and third terms. Thus we obtain

$$A_{\mu\sigma\tau} - A_{\mu\tau\sigma} = B^{\rho}_{\mu\sigma\tau} A_{\rho} \quad . \quad . \quad . \quad (42)$$

where

$$B^{\rho}_{\mu\sigma\tau} = -\frac{\partial}{\partial x_{\tau}}\{\mu\sigma, \rho\} + \frac{\partial}{\partial x_{\sigma}}\{\mu\tau, \rho\} - \{\mu\sigma, \alpha\}\{\alpha\tau, \rho\}$$
$$+ \{\mu\tau, \alpha\}\{\alpha\sigma, \rho\} \quad (43)$$

The essential feature of the result is that on the right side of (42) the A_{ρ} occur alone, without their derivatives. From the tensor character of $A_{\mu\sigma\tau} - A_{\mu\tau\sigma}$ in conjunction with the fact that A_{ρ} is an arbitrary vector, it follows, by reason of § 7, that $B^{\rho}_{\mu\sigma\tau}$ is a tensor (the Riemann-Christoffel tensor).

The mathematical importance of this tensor is as follows: If the continuum is of such a nature that there is a co-ordinate system with reference to which the $g_{\mu\nu}$ are constants, then all the $B^{\rho}_{\mu\sigma\tau}$ vanish. If we choose any new system of co-ordinates in place of the original ones, the $g_{\mu\nu}$ referred thereto will not be constants, but in consequence of its tensor nature, the transformed components of $B^{\rho}_{\mu\sigma\tau}$ will still vanish in the new system. Thus the vanishing of the Riemann tensor is a necessary condition that, by an appropriate choice of the system of reference, the $g_{\mu\nu}$ may be constants. In our problem this corresponds to the case in which,* with a suitable choice of the system of reference, the special theory of relativity holds good for a *finite* region of the continuum.

Contracting (43) with respect to the indices τ and ρ we obtain the covariant tensor of second rank

* The mathematicians have proved that this is also a *sufficient* condition.

$$\left.\begin{aligned} G_{\mu\nu} &= B^{\rho}_{\mu\nu\rho} = R_{\mu\nu} + S_{\mu\nu} \\ \text{where} \quad & \\ R_{\mu\nu} &= -\frac{\partial}{\partial x_{\alpha}}\{\mu\nu, \alpha\} + \{\mu\alpha, \beta\}\{\nu\beta, \alpha\} \\ S_{\mu\nu} &= \frac{\partial^2 \log\sqrt{-g}}{\partial x_{\mu}\partial x_{\nu}} - \{\mu\nu, \alpha\}\frac{\partial \log\sqrt{-g}}{\partial x_{\alpha}} \end{aligned}\right\} \quad (44)$$

Note on the Choice of Co-ordinates.—It has already been observed in § 8, in connexion with equation (18a), that the choice of co-ordinates may with advantage be made so that $\sqrt{-g} = 1$. A glance at the equations obtained in the last two sections shows that by such a choice the laws of formation of tensors undergo an important simplification. This applies particularly to $G_{\mu\nu}$, the tensor just developed, which plays a fundamental part in the theory to be set forth. For this specialization of the choice of co-ordinates brings about the vanishing of $S_{\mu\nu}$, so that the tensor $G_{\mu\nu}$ reduces to $R_{\mu\nu}$.

On this account I shall hereafter give all relations in the simplified form which this specialization of the choice of co-ordinates brings with it. It will then be an easy matter to revert to the *generally* covariant equations, if this seems desirable in a special case.

C. Theory of the Gravitational Field

§ 13. Equations of Motion of a Material Point in the Gravitational Field. Expression for the Field-components of Gravitation

A freely movable body not subjected to external forces moves, according to the special theory of relativity, in a straight line and uniformly. This is also the case, according to the general theory of relativity, for a part of four-dimensional space in which the system of co-ordinates K_0, may be, and is, so chosen that they have the special constant values given in (4).

If we consider precisely this movement from any chosen system of co-ordinates K_1, the body, observed from K_1, moves, according to the considerations in § 2, in a gravitational field. The law of motion with respect to K_1 results without difficulty from the following consideration. With respect to K_0 the law of motion corresponds to a four-dimensional straight line, i.e. to a geodetic line. Now since the geodetic line is defined independently of the system of reference, its equations will also be the equation of motion of the material point with respect to K_1. If we set

$$\Gamma^{\tau}_{\mu\nu} = - \{\mu\nu, \tau\} \quad . \quad . \quad . \quad (45)$$

the equation of the motion of the point with respect to K_1, becomes

$$\frac{d^2 x_\tau}{ds^2} = \Gamma^{\tau}_{\mu\nu} \frac{dx_\mu}{ds} \frac{dx_\nu}{ds} \quad . \quad . \quad . \quad (46)$$

We now make the assumption, which readily suggests itself, that this covariant system of equations also defines the motion of the point in the gravitational field in the case when there is no system of reference K_0, with respect to which the special theory of relativity holds good in a finite region. We have all the more justification for this assumption as (46) contains only *first* derivatives of the $g_{\mu\nu}$, between which even in the special case of the existence of K_0, no relations subsist.*

If the $\Gamma^{\tau}_{\mu\nu}$ vanish, then the point moves uniformly in a straight line. These quantities therefore condition the deviation of the motion from uniformity. They are the components of the gravitational field.

§ 14. The Field Equations of Gravitation in the Absence of Matter

We make a distinction hereafter between " gravitational field " and " matter " in this way, that we denote everything but the gravitational field as " matter." Our use of the word therefore includes not only matter in the ordinary sense, but the electromagnetic field as well.

Our next task is to find the field equations of gravitation in the absence of matter. Here we again apply the method

* It is only between the second (and first) derivatives that, by § 12, the relations $B^{\rho}_{\mu\sigma\tau} = 0$ subsist.

employed in the preceding paragraph in formulating the equations of motion of the material point. A special case in which the required equations must in any case be satisfied is that of the special theory of relativity, in which the $g_{\mu\nu}$ have certain constant values. Let this be the case in a certain finite space in relation to a definite system of co-ordinates K_0. Relatively to this system all the components of the Riemann tensor $B^{\rho}_{\mu\sigma\tau}$, defined in (43), vanish. For the space under consideration they then vanish, also in any other system of co-ordinates.

Thus the required equations of the matter-free gravitational field must in any case be satisfied if all $B^{\rho}_{\mu\sigma\tau}$ vanish. But this condition goes too far. For it is clear that, e.g., the gravitational field generated by a material point in its environment certainly cannot be "transformed away" by any choice of the system of co-ordinates, i.e. it cannot be transformed to the case of constant $g_{\mu\nu}$.

This prompts us to require for the matter-free gravitational field that the symmetrical tensor $G_{\mu\nu}$, derived from the tensor $B^{\rho}_{\mu\nu\tau}$, shall vanish. Thus we obtain ten equations for the ten quantities $g_{\mu\nu}$, which are satisfied in the special case of the vanishing of all $B^{\rho}_{\mu\nu\tau}$. With the choice which we have made of a system of co-ordinates, and taking (44) into consideration, the equations for the matter-free field are

$$\left.\begin{aligned} \frac{\partial \Gamma^{\alpha}_{\mu\nu}}{\partial x_{\alpha}} + \Gamma^{\alpha}_{\mu\beta}\Gamma^{\beta}_{\nu\alpha} &= 0 \\ \sqrt{-g} &= 1 \end{aligned}\right\} \quad . \quad . \quad . \quad (47)$$

It must be pointed out that there is only a minimum of arbitrariness in the choice of these equations. For besides $G_{\mu\nu}$ there is no tensor of second rank which is formed from the $g_{\mu\nu}$ and its derivatives, contains no derivations higher than second, and is linear in these derivatives.*

These equations, which proceed, by the method of pure

* Properly speaking, this can be affirmed only of the tensor

$$G_{\mu\nu} + \lambda g_{\mu\nu} g^{\alpha\beta} G_{\alpha\beta},$$

where λ is a constant. If, however, we set this tensor = 0, we come back again to the equations $G_{\mu\nu} = 0$.

mathematics, from the requirement of the general theory of relativity, give us, in combination with the equations of motion (46), to a first approximation Newton's law of attraction, and to a second approximation the explanation of the motion of the perihelion of the planet Mercury discovered by Leverrier (as it remains after corrections for perturbation have been made). These facts must, in my opinion, be taken as a convincing proof of the correctness of the theory.

§ 15. The Hamiltonian Function for the Gravitational Field. Laws of Momentum and Energy

To show that the field equations correspond to the laws of momentum and energy, it is most convenient to write them in the following Hamiltonian form :—

$$\left.\begin{aligned} \delta\int \mathrm{H}d\tau &= 0 \\ \mathrm{H} &= g^{\mu\nu}\Gamma^{\alpha}_{\mu\beta}\Gamma^{\beta}_{\nu\alpha} \\ \sqrt{-g} &= 1 \end{aligned}\right\} \quad . \quad . \quad . \quad (47\mathrm{a})$$

where, on the boundary of the finite four-dimensional region of integration which we have in view, the variations vanish.

We first have to show that the form (47a) is equivalent to the equations (47). For this purpose we regard H as a function of the $g^{\mu\nu}$ and the $g^{\mu\nu}_{\sigma}$ $(= \partial g^{\mu\nu}/\partial x_\sigma)$.

Then in the first place

$$\begin{aligned} \delta\mathrm{H} &= \Gamma^{\alpha}_{\mu\beta}\Gamma^{\beta}_{\nu\alpha}\,\delta g^{\mu\nu} + 2g^{\mu\nu}\Gamma^{\alpha}_{\mu\beta}\,\delta\Gamma^{\beta}_{\nu\alpha} \\ &= -\,\Gamma^{\alpha}_{\mu\beta}\Gamma^{\beta}_{\nu\alpha}\,\delta g^{\mu\nu} + 2\Gamma^{\alpha}_{\mu\beta}\,\delta(g^{\mu\nu}\Gamma^{\beta}_{\nu\alpha}). \end{aligned}$$

But

$$\delta\left(g^{\mu\nu}\Gamma^{\beta}_{\nu\alpha}\right) = -\tfrac{1}{2}\delta\left[g^{\mu\nu}g^{\beta\lambda}\left(\frac{\partial g_{\nu\lambda}}{\partial x_\alpha} + \frac{\partial g_{\alpha\lambda}}{\partial x_\nu} - \frac{\partial g_{\alpha\nu}}{\partial x_\lambda}\right)\right].$$

The terms arising from the last two terms in round brackets are of different sign, and result from each other (since the denomination of the summation indices is immaterial) through interchange of the indices μ and β. They cancel each other in the expression for δH, because they are multiplied by the quantity $\Gamma^{\alpha}_{\mu\beta}$, which is symmetrical with respect to the indices μ and β. Thus there remains only the first term in round brackets to be considered, so that, taking (31) into account, we obtain

$$\delta H = -\Gamma^{\alpha}_{\mu\beta}\Gamma^{\beta}_{\nu\alpha}\delta g^{\mu\nu} + \Gamma^{\alpha}_{\mu\beta}\delta g^{\mu\beta}_{\alpha}.$$

Thus

$$\left.\begin{aligned} \frac{\partial H}{\partial g^{\mu\nu}} &= -\Gamma^{\alpha}_{\mu\beta}\Gamma^{\beta}_{\nu\alpha} \\ \frac{\partial H}{\partial g^{\mu\nu}_{\sigma}} &= \Gamma^{\sigma}_{\mu\nu} \end{aligned}\right\} \quad . \quad . \quad . \quad (48)$$

Carrying out the variation in (47a), we get in the first place

$$\frac{\partial}{\partial x_{\alpha}}\left(\frac{\partial H}{\partial g^{\mu\nu}_{\alpha}}\right) - \frac{\partial H}{\partial g^{\mu\nu}} = 0, \quad . \quad . \quad . \quad (47b)$$

which, on account of (48), agrees with (47), as was to be proved.

If we multiply (47b) by $g^{\mu\nu}_{\sigma}$, then because

$$\frac{\partial g^{\mu\nu}_{\sigma}}{\partial x_{\alpha}} = \frac{\partial g^{\mu\nu}_{\alpha}}{\partial x_{\sigma}}$$

and, consequently,

$$g^{\mu\nu}_{\sigma}\frac{\partial}{\partial x_{\alpha}}\left(\frac{\partial H}{\partial g^{\mu\nu}_{\alpha}}\right) = \frac{\partial}{\partial x_{\alpha}}\left(g^{\mu\nu}_{\sigma}\frac{\partial H}{\partial g^{\mu\nu}_{\alpha}}\right) - \frac{\partial H}{\partial g^{\mu\nu}_{\alpha}}\frac{\partial g^{\mu\nu}_{\alpha}}{\partial x_{\sigma}},$$

we obtain the equation

$$\frac{\partial}{\partial x_{\alpha}}\left(g^{\mu\nu}_{\sigma}\frac{\partial H}{\partial g^{\mu\nu}_{\alpha}}\right) - \frac{\partial H}{\partial x_{\sigma}} = 0$$

or *

$$\left.\begin{aligned} \frac{\partial t^{\alpha}_{\sigma}}{\partial x_{\alpha}} &= 0 \\ -2\kappa t^{\alpha}_{\sigma} &= g^{\mu\nu}_{\sigma}\frac{\partial H}{\partial g^{\mu\nu}_{\alpha}} - \delta^{\alpha}_{\sigma}H \end{aligned}\right\} \quad . \quad . \quad . \quad (49)$$

where, on account of (48), the second equation of (47), and (34)

$$\kappa t^{\alpha}_{\sigma} = \tfrac{1}{2}\delta^{\alpha}_{\sigma}g^{\mu\nu}\Gamma^{\lambda}_{\mu\beta}\Gamma^{\beta}_{\nu\lambda} - g^{\mu\nu}\Gamma^{\alpha}_{\mu\beta}\Gamma^{\beta}_{\nu\sigma} \quad . \quad . \quad (50)$$

* The reason for the introduction of the factor -2κ will be apparent later.

It is to be noticed that t_σ^α is not a tensor; on the other hand (49) applies to all systems of co-ordinates for which $\sqrt{-g} = 1$. This equation expresses the law of conservation of momentum and of energy for the gravitational field. Actually the integration of this equation over a three-dimensional volume V yields the four equations

$$\frac{d}{dx_4}\int t_\sigma^4 dV = \int (lt_\sigma^1 + mt_\sigma^2 + nt_\sigma^3) dS \quad . \quad . \quad (49a)$$

where l, m, n denote the direction-cosines of direction of the inward drawn normal at the element dS of the bounding surface (in the sense of Euclidean geometry). We recognize in this the expression of the laws of conservation in their usual form. The quantities t_σ^α we call the "energy components" of the gravitational field.

I will now give equations (47) in a third form, which is particularly useful for a vivid grasp of our subject. By multiplication of the field equations (47) by $g^{\nu\sigma}$ these are obtained in the "mixed" form. Note that

$$g^{\nu\sigma}\frac{\partial \Gamma_{\mu\nu}^\alpha}{\partial x_\alpha} = \frac{\partial}{\partial x_\alpha}\left(g^{\nu\sigma}\Gamma_{\mu\nu}^\alpha\right) - \frac{\partial g^{\nu\sigma}}{\partial x_\alpha}\Gamma_{\mu\nu}^\alpha,$$

which quantity, by reason of (34), is equal to

$$\frac{\partial}{\partial x_\alpha}\left(g^{\nu\sigma}\Gamma_{\mu\nu}^\alpha\right) - g^{\nu\beta}\Gamma_{\alpha\beta}^\sigma\Gamma_{\mu\nu}^\alpha - g^{\sigma\beta}\Gamma_{\beta\alpha}^\nu\Gamma_{\mu\nu}^\alpha,$$

or (with different symbols for the summation indices)

$$\frac{\partial}{\partial x_\alpha}\left(g^{\sigma\beta}\Gamma_{\mu\beta}^\alpha\right) - g^{\gamma\delta}\Gamma_{\gamma\beta}^\sigma\Gamma_{\delta\mu}^\beta - g^{\nu\sigma}\Gamma_{\mu\beta}^\alpha\Gamma_{\nu\alpha}^\beta.$$

The third term of this expression cancels with the one arising from the second term of the field equations (47); using relation (50), the second term may be written

$$\kappa(t_\mu^\sigma - \tfrac{1}{2}\delta_\mu^\sigma t),$$

where $t = t_\alpha^\alpha$. Thus instead of equations (47) we obtain

$$\left.\begin{aligned}\frac{\partial}{\partial x_\alpha}\left(g^{\sigma\beta}\Gamma_{\mu\beta}^\alpha\right) &= -\kappa(t_\mu^\sigma - \tfrac{1}{2}\delta_\mu^\sigma t)\\ \sqrt{-g} &= 1\end{aligned}\right\} \quad . \quad . \quad (51)$$

§ 16. The General Form of the Field Equations of Gravitation

The field equations for matter-free space formulated in § 15 are to be compared with the field equation

$$\nabla^2\phi = 0$$

of Newton's theory. We require the equation corresponding to Poisson's equation

$$\nabla^2\phi = 4\pi\kappa\rho,$$

where ρ denotes the density of matter.

The special theory of relativity has led to the conclusion that inert mass is nothing more or less than energy, which finds its complete mathematical expression in a symmetrical tensor of second rank, the energy-tensor. Thus in the general theory of relativity we must introduce a corresponding energy-tensor of matter T_σ^α, which, like the energy-components t_σ [equations (49) and (50)] of the gravitational field, will have mixed character, but will pertain to a symmetrical covariant tensor.*

The system of equation (51) shows how this energy-tensor (corresponding to the density ρ in Poisson's equation) is to be introduced into the field equations of gravitation. For if we consider a complete system (e.g. the solar system), the total mass of the system, and therefore its total gravitating action as well, will depend on the total energy of the system, and therefore on the ponderable energy together with the gravitational energy. This will allow itself to be expressed by introducing into (51), in place of the energy-components of the gravitational field alone, the sums $t_\mu^\sigma + T_\mu^\sigma$ of the energy-components of matter and of gravitational field. Thus instead of (51) we obtain the tensor equation

$$\left.\begin{aligned} \frac{\partial}{\partial x_\alpha}(g^{\sigma\beta}\Gamma_{\mu\beta}^\alpha) &= -\kappa[(t_\mu^\sigma + T_\mu^\sigma) - \tfrac{1}{2}\delta_\mu^\sigma(t + T)], \\ \sqrt{-g} &= 1 \end{aligned}\right\} . \quad (52)$$

where we have set $T = T_\mu^\mu$ (Laue's scalar). These are the

* $g_{\alpha\tau}T_\sigma^\alpha = T_{\sigma\tau}$ and $g^{\sigma\beta}T_\sigma^\alpha = T^{\alpha\beta}$ are to be symmetrical tensors.

required general field equations of gravitation in mixed form. Working back from these, we have in place of (47)

$$\left.\begin{aligned}\frac{\partial}{\partial x_\alpha}\Gamma^\alpha_{\mu\nu} + \Gamma^\alpha_{\mu\beta}\Gamma^\beta_{\nu\alpha} &= -\kappa(T_{\mu\nu} - \tfrac{1}{2}g_{\mu\nu}T),\\ \sqrt{-g} &= 1\end{aligned}\right\} \quad . \quad (53)$$

It must be admitted that this introduction of the energy-tensor of matter is not justified by the relativity postulate alone. For this reason we have here deduced it from the requirement that the energy of the gravitational field shall act gravitatively in the same way as any other kind of energy. But the strongest reason for the choice of these equations lies in their consequence, that the equations of conservation of momentum and energy, corresponding exactly to equations (49) and (49a), hold good for the components of the total energy. This will be shown in § 17.

§ 17. The Laws of Conservation in the General Case

Equation (52) may readily be transformed so that the second term on the right-hand side vanishes. Contract (52) with respect to the indices μ and σ, and after multiplying the resulting equation by $\frac{1}{2}\delta^\sigma_\mu$, subtract it from equation (52). This gives

$$\frac{\partial}{\partial x_\alpha}(g^{\sigma\beta}\Gamma^\alpha_{\mu\beta} - \tfrac{1}{2}\delta^\sigma_\mu g^{\lambda\beta}\Gamma^\alpha_{\lambda\beta}) = -\kappa(t^\sigma_\mu + T^\sigma_\mu). \qquad (52a)$$

On this equation we perform the operation $\partial/\partial x_\sigma$. We have

$$\frac{\partial^2}{\partial x_\alpha \partial x_\sigma}\left(g^\sigma \Gamma^{\ \alpha}_{\beta\mu}\right) = -\tfrac{1}{2}\frac{\partial^2}{\partial x_\alpha \partial x_\sigma}\left[g^{\sigma\beta}g^{\alpha\lambda}\left(\frac{\partial g_{\mu\lambda}}{\partial x_\beta} + \frac{\partial g_{\beta\lambda}}{\partial x_\mu} - \frac{\partial g_{\mu\beta}}{\partial x_\lambda}\right)\right].$$

The first and third terms of the round brackets yield contributions which cancel one another, as may be seen by interchanging, in the contribution of the third term, the summation indices α and σ on the one hand, and β and λ on the other. The second term may be re-modelled by (31), so that we have

$$\frac{\partial^2}{\partial x_\alpha \partial x_\sigma}\left(g^{\sigma\beta}\Gamma^\alpha_{\mu\beta}\right) = \tfrac{1}{2}\frac{\partial^3 g^{\alpha\beta}}{\partial x_\alpha \partial x_\beta \partial x_\mu} \quad . \quad . \quad (54)$$

The second term on the left-hand side of (52a) yields in the first place

$$-\tfrac{1}{2}\frac{\partial^2}{\partial x_\alpha \partial x_\mu}\left(g^{\lambda\beta}\Gamma^\alpha_{\lambda\beta}\right)$$

or

$$\tfrac{1}{4}\frac{\partial^2}{\partial x_\alpha \partial x_\mu}\left[g^{\lambda\beta}g^{\alpha\delta}\left(\frac{\partial g_{\delta\lambda}}{\partial x_\beta}+\frac{\partial g_{\delta\beta}}{\partial x_\lambda}-\frac{\partial g_{\lambda\beta}}{\partial x_\delta}\right)\right].$$

With the choice of co-ordinates which we have made, the term deriving from the last term in round brackets disappears by reason of (29). The other two may be combined, and together, by (31), they give

$$-\tfrac{1}{2}\frac{\partial^3 g^{\alpha\beta}}{\partial x_\alpha \partial x_\beta \partial x_\mu},$$

so that in consideration of (54), we have the identity

$$\frac{\partial^2}{\partial x_\alpha \partial x_\sigma}\left(g^{\rho\beta}\Gamma_{\mu\beta}-\tfrac{1}{2}\delta^\sigma_\mu g^{\lambda\beta}\Gamma^\alpha_{\lambda\beta}\right)\equiv 0 \quad . \quad . \quad (55)$$

From (55) and (52a), it follows that

$$\frac{\partial(t^\sigma_\mu + \mathrm{T}^\sigma_\mu)}{\partial x_\sigma} = 0. \quad . \quad . \quad . \quad (56)$$

Thus it results from our field equations of gravitation that the laws of conservation of momentum and energy are satisfied. This may be seen most easily from the consideration which leads to equation (49a); except that here, instead of the energy components t^σ of the gravitational field, we have to introduce the totality of the energy components of matter and gravitational field.

§ 18. The Laws of Momentum and Energy for Matter, as a Consequence of the Field Equations

Multiplying (53) by $\partial g^{\mu\nu}/\partial x_\sigma$, we obtain, by the method adopted in § 15, in view of the vanishing of

$$g_{\mu\nu}\frac{\partial g^{\mu\nu}}{\partial x_\sigma},$$

the equation

$$\frac{\partial t^\alpha_\sigma}{\partial x_\alpha} + \tfrac{1}{2}\frac{\partial g^{\mu\nu}}{\partial x_\sigma}\mathrm{T}_{\mu\nu} = 0,$$

or, in view of (56),

$$\frac{\partial T^{\alpha}_{\sigma}}{\partial x_{\alpha}} + \frac{1}{2}\frac{\partial g^{\mu\nu}}{\partial x_{\sigma}}T_{\mu\nu} = 0 \quad . \quad . \quad . \quad (57)$$

Comparison with (41b) shows that with the choice of system of co-ordinates which we have made, this equation predicates nothing more or less than the vanishing of divergence of the material energy-tensor. Physically, the occurrence of the second term on the left-hand side shows that laws of conservation of momentum and energy do not apply in the strict sense for matter alone, or else that they apply only when the $g^{\mu\nu}$ are constant, i.e. when the field intensities of gravitation vanish. This second term is an expression for momentum, and for energy, as transferred per unit of volume and time from the gravitational field to matter. This is brought out still more clearly by re-writing (57) in the sense of (41) as

$$\frac{\partial T^{\alpha}_{\sigma}}{\partial x_{\alpha}} = -\Gamma^{\beta}_{\alpha\sigma}T^{\alpha}_{\beta} \quad . \quad . \quad . \quad (57a)$$

The right side expresses the energetic effect of the gravitational field on matter.

Thus the field equations of gravitation contain four conditions which govern the course of material phenomena. They give the equations of material phenomena completely, if the latter is capable of being characterized by four differential equations independent of one another.*

D. Material Phenomena

The mathematical aids developed in part B enable us forthwith to generalize the physical laws of matter (hydrodynamics, Maxwell's electrodynamics), as they are formulated in the special theory of relativity, so that they will fit in with the general theory of relativity. When this is done, the general principle of relativity does not indeed afford us a further limitation of possibilities ; but it makes us acquainted with the influence of the gravitational field on all processes,

* On this question cf. H. Hilbert, Nachr. d. K. Gesellsch. d. Wiss. zu Göttingen, Math.-phys. Klasse, 1915, p. 3.

without our having to introduce any new hypothesis whatever.

Hence it comes about that it is not necessary to introduce definite assumptions as to the physical nature of matter (in the narrower sense). In particular it may remain an open question whether the theory of the electromagnetic field in conjunction with that of the gravitational field furnishes a sufficient basis for the theory of matter or not. The general postulate of relativity is unable on principle to tell us anything about this. It must remain to be seen, during the working out of the theory, whether electromagnetics and the doctrine of gravitation are able in collaboration to perform what the former by itself is unable to do.

§ 19. Euler's Equations for a Frictionless Adiabatic Fluid

Let p and ρ be two scalars, the former of which we call the "pressure," the latter the "density" of a fluid; and let an equation subsist between them. Let the contravariant symmetrical tensor

$$T^{\alpha\beta} = -g^{\alpha\beta}p + \rho\frac{dx_\alpha}{ds}\frac{dx_\beta}{ds} \quad . \quad . \quad . \quad (58)$$

be the contravariant energy-tensor of the fluid. To it belongs the covariant tensor

$$T_{\mu\nu} = -g_{\mu\nu}p + g_{\mu\alpha}g_{\mu\beta}\frac{dx_\alpha}{ds}\frac{dx_\beta}{ds}\rho, \quad . \quad . \quad (58a)$$

as well as the mixed tensor *

$$T^\alpha_\sigma = -\delta^\alpha_\sigma p + g_{\sigma\beta}\frac{dx_\beta}{ds}\frac{dx_\alpha}{ds}\rho \quad . \quad . \quad (58b)$$

Inserting the right-hand side of (58b) in (57a), we obtain the Eulerian hydrodynamical equations of the general theory of relativity. They give, in theory, a complete solution of the problem of motion, since the four equations (57a), together

* For an observer using a system of reference in the sense of the special theory of relativity for an infinitely small region, and moving with it, the density of energy T^4_4 equals $\rho - p$. This gives the definition of ρ. Thus ρ is not constant for an incompressible fluid.

with the given equation between p and ρ, and the equation

$$g_{\alpha\beta}\frac{dx_\alpha}{ds}\frac{dx_\beta}{ds} = 1,$$

are sufficient, $g_{\alpha\beta}$ being given, to define the six unknowns

$$p, \rho, \frac{dx_1}{ds}, \frac{dx_2}{ds}, \frac{dx_3}{ds}, \frac{dx_4}{ds}.$$

If the $g_{\mu\nu}$ are also unknown, the equations (53) are brought in. These are eleven equations for defining the ten functions $g_{\mu\nu}$, so that these functions appear over-defined. We must remember, however, that the equations (57a) are already contained in the equations (53), so that the latter represent only seven independent equations. There is good reason for this lack of definition, in that the wide freedom of the choice of co-ordinates causes the problem to remain mathematically undefined to such a degree that three of the functions of space may be chosen at will.*

§ 20. Maxwell's Electromagnetic Field Equations for Free Space

Let ϕ_ν be the components of a covariant vector—the electromagnetic potential vector. From them we form, in accordance with (36), the components $F_{\rho\sigma}$ of the covariant six-vector of the electromagnetic field, in accordance with the system of equations

$$F_{\rho\sigma} = \frac{\partial\phi_\rho}{\partial x_\sigma} - \frac{\partial\phi_\sigma}{\partial x_\rho} \quad . \quad . \quad . \quad (59)$$

It follows from (59) that the system of equations

$$\frac{\partial F_{\rho\sigma}}{\partial x_\tau} + \frac{\partial F_{\sigma\tau}}{\partial x_\rho} + \frac{\partial F_{\tau\rho}}{\partial x_\sigma} = 0 \quad . \quad . \quad . \quad (60)$$

is satisfied, its left side being, by (37), an antisymmetrical tensor of the third rank. System (60) thus contains essentially four equations which are written out as follows:—

* On the abandonment of the choice of co-ordinates with $g = -1$, there remain *four* functions of space with liberty of choice, corresponding to the four arbitrary functions at our disposal in the choice of co-ordinates.

$$\left.\begin{aligned}\frac{\partial F_{23}}{\partial x_4}+\frac{\partial F_{34}}{\partial x_2}+\frac{\partial F_{42}}{\partial x_3}=0\\ \frac{\partial F_{34}}{\partial x_1}+\frac{\partial F_{41}}{\partial x_3}+\frac{\partial F_{13}}{\partial x_4}=0\\ \frac{\partial F_{41}}{\partial x_2}+\frac{\partial F_{12}}{\partial x_4}+\frac{\partial F_{24}}{\partial x_1}=0\\ \frac{\partial F_{12}}{\partial x_3}+\frac{\partial F_{23}}{\partial x_1}+\frac{\partial F_{31}}{\partial x_2}=0\end{aligned}\right\} \quad . \quad . \quad (60a)$$

This system corresponds to the second of Maxwell's systems of equations. We recognize this at once by setting

$$\left.\begin{aligned}F_{23}=H_x,\ F_{14}=E_x\\ F_{31}=H_y,\ F_{24}=E_y\\ F_{12}=H_z,\ F_{34}=E_z\end{aligned}\right\} \quad . \quad . \quad . \quad (61)$$

Then in place of (60a) we may set, in the usual notation of three-dimensional vector analysis,

$$\left.\begin{aligned}-\frac{\partial H}{\partial t}=\text{curl } E\\ \text{div } H=0\end{aligned}\right\} \quad . \quad . \quad . \quad (60b)$$

We obtain Maxwell's first system by generalizing the form given by Minkowski. We introduce the contravariant six-vector associated with $F^{\alpha\beta}$

$$F^{\mu\nu}=g^{\mu\alpha}g^{\nu\beta}F_{\alpha\beta} \quad . \quad . \quad . \quad (62)$$

and also the contravariant vector J^{μ} of the density of the electric current. Then, taking (40) into consideration, the following equations will be invariant for any substitution whose invariant is unity (in agreement with the chosen co-ordinates) :—

$$\frac{\partial}{\partial x_\nu}F^{\mu\nu}=J^{\mu} \quad . \quad . \quad . \quad . \quad (63)$$

Let

$$\left.\begin{aligned}F^{23}=H'_x,\ F^{14}=-E'_x\\ F^{31}=H'_y,\ F^{24}=-E'_y\\ F^{12}=H'_z,\ F^{34}=-E'_z\end{aligned}\right\} \quad . \quad . \quad (64)$$

which quantities are equal to the quantities H_x . . . E^z in the special case of the restricted theory of relativity; and in addition

$$J^1 = j_x,\ J^2 = j_y,\ J^3 = j_z,\ J^4 = \rho,$$

we obtain in place of (63)

$$\left.\begin{aligned}\frac{\partial E'}{\partial t} + j &= \text{curl } H' \\ \text{div } E' &= \rho\end{aligned}\right\} \quad . \quad . \quad . \quad (63a)$$

The equations (60), (62), and (63) thus form the generalization of Maxwell's field equations for free space, with the convention which we have established with respect to the choice of co-ordinates.

The Energy-components of the Electromagnetic Field.—We form the inner product

$$\kappa_\sigma = F_{\sigma\mu}J^\mu \quad . \quad . \quad . \quad . \quad (65)$$

By (61) its components, written in the three-dimensional manner, are

$$\left.\begin{aligned}\kappa_1 &= \rho E_x + [j \,.\, H]^x \\ &\cdot \quad \cdot \quad \cdot \quad \cdot \\ &\cdot \quad \cdot \quad \cdot \quad \cdot \\ \kappa_4 &= -(jE)\end{aligned}\right\} \quad . \quad . \quad . \quad (65a)$$

κ_σ is a covariant vector the components of which are equal to the negative momentum, or, respectively, the energy, which is transferred from the electric masses to the electromagnetic field per unit of time and volume. If the electric masses are free, that is, under the sole influence of the electromagnetic field, the covariant vector κ_σ will vanish.

To obtain the energy-components T_σ^ν of the electromagnetic field, we need only give to equation $\kappa_\sigma = 0$ the form of equation (57). From (63) and (65) we have in the first place

$$\kappa_\sigma = F_{\sigma\mu}\frac{\partial F^{\mu\nu}}{\partial x_\nu} = \frac{\partial}{\partial x_\nu}(F_{\sigma\mu}F^{\mu\nu}) - F^{\mu\nu}\frac{\partial F_{\sigma\mu}}{\partial x_\nu}.$$

The second term of the right-hand side, by reason of (60), permits the transformation

$$F^{\mu\nu}\frac{\partial F_{\sigma\mu}}{\partial x_\nu} = -\tfrac{1}{2}F^{\mu\nu}\frac{\partial F_{\mu\nu}}{\partial x_\sigma} = -\tfrac{1}{2}g^{\mu\alpha}g^{\nu\beta}F_{\alpha\beta}\frac{\partial F_{\mu\nu}}{\partial x_\sigma},$$

which latter expression may, for reasons of symmetry, also be written in the form

$$-\tfrac{1}{4}\left[g^{\mu\alpha}g^{\nu\beta}F_{\alpha\beta}\frac{\partial F_{\mu\nu}}{\partial x_\sigma} + g^{\mu\alpha}g^{\nu\beta}\frac{\partial F_{\alpha\beta}}{\partial x_\sigma}F_{\mu\nu}\right].$$

But for this we may set

$$-\tfrac{1}{4}\frac{\partial}{\partial x_\sigma}(g^{\mu\alpha}g^{\nu\beta}F_{\alpha\beta}F_{\mu\nu}) + \tfrac{1}{4}F_{\alpha\beta}F_{\mu\nu}\frac{\partial}{\partial x_\sigma}(g^{\mu\alpha}g^{\nu\beta}).$$

The first of these terms is written more briefly

$$-\tfrac{1}{4}\frac{\partial}{\partial x_\sigma}(F^{\mu\nu}F_{\mu\nu});$$

the second, after the differentiation is carried out, and after some reduction, results in

$$-\tfrac{1}{2}F^{\mu\tau}F_{\mu\nu}g^{\nu\rho}\frac{\partial g_{\sigma\tau}}{\partial x_\sigma}.$$

Taking all three terms together we obtain the relation

$$\kappa_\sigma = \frac{\partial T_\sigma^\nu}{\partial x_\nu} - \tfrac{1}{2}g^{\tau\mu}\frac{\partial g_{\mu\nu}}{\partial x_\sigma}T_\tau^\nu \quad . \quad . \quad . \quad (66)$$

where

$$T_\sigma^\nu = -F_{\sigma\alpha}F^{\nu\alpha} + \tfrac{1}{4}\delta_\sigma^\nu F_{\alpha\beta}F^{\alpha\beta}.$$

Equation (66), if κ_σ vanishes, is, on account of (30), equivalent to (57) or (57a) respectively. Therefore the T_σ^ν are the energy-components of the electromagnetic field. With the help of (61) and (64), it is easy to show that these energy-components of the electromagnetic field in the case of the special theory of relativity give the well-known Maxwell-Poynting expressions.

We have now deduced the general laws which are satisfied by the gravitational field and matter, by consistently using a system of co-ordinates for which $\sqrt{-g} = 1$. We have thereby achieved a considerable simplification of formulæ and calculations, without failing to comply with the requirement of general covariance; for we have drawn our equations from generally covariant equations by specializing the system of co-ordinates.

Still the question is not without a formal interest, whether with a correspondingly generalized definition of the energy-components of gravitational field and matter, even without specializing the system of co-ordinates, it is possible to formulate laws of conservation in the form of equation (56), and field equations of gravitation of the same nature as (52) or (52a), in such a manner that on the left we have a divergence (in the ordinary sense), and on the right the sum of the energy-components of matter and gravitation. I have found that in both cases this is actually so. But I do not think that the communication of my somewhat extensive reflexions on this subject would be worth while, because after all they do not give us anything that is materially new.

E

§ 21. Newton's Theory as a First Approximation

As has already been mentioned more than once, the special theory of relativity as a special case of the general theory is characterized by the $g_{\mu\nu}$ having the constant values (4). From what has already been said, this means complete neglect of the effects of gravitation. We arrive at a closer approximation to reality by considering the case where the $g_{\mu\nu}$ differ from the values of (4) by quantities which are small compared with 1, and neglecting small quantities of second and higher order. (First point of view of approximation.)

It is further to be assumed that in the space-time territory under consideration the $g_{\mu\nu}$ at spatial infinity, with a suitable choice of co-ordinates, tend toward the values (4); i.e. we are considering gravitational fields which may be regarded as generated exclusively by matter in the finite region.

It might be thought that these approximations must lead us to Newton's theory. But to that end we still need to approximate the fundamental equations from a second point of view. We give our attention to the motion of a material point in accordance with the equations (16). In the case of the special theory of relativity the components

$$\frac{dx_1}{ds}, \frac{dx_2}{ds}, \frac{dx_3}{ds}$$

may take on any values. This signifies that any velocity

$$v = \sqrt{\left(\frac{dx_1}{dx_4}\right)^2 + \left(\frac{dx_2}{dx_4}\right)^2 + \left(\frac{dx_3}{dx_4}\right)^2}$$

may occur, which is less than the velocity of light *in vacuo*. If we restrict ourselves to the case which almost exclusively offers itself to our experience, of v being small as compared with the velocity of light, this denotes that the components

$$\frac{dx_1}{ds}, \frac{dx_2}{ds}, \frac{dx_3}{ds}$$

are to be treated as small quantities, while dx_4/ds, to the second order of small quantities, is equal to one. (Second point of view of approximation.)

Now we remark that from the first point of view of approximation the magnitudes $\Gamma^{\tau}_{\mu\nu}$ are all small magnitudes of at least the first order. A glance at (46) thus shows that in this equation, from the second point of view of approximation, we have to consider only terms for which $\mu = \nu = 4$. Restricting ourselves to terms of lowest order we first obtain in place of (46) the equations

$$\frac{d^2x_\tau}{dt^2} = \Gamma^{\tau}_{44}$$

where we have set $ds = dx_4 = dt$; or with restriction to terms which from the first point of view of approximation are of first order :—

$$\frac{d^2x_\tau}{dt^2} = [44, \tau] \quad (\tau = 1, 2, 3)$$

$$\frac{d^2x_4}{dt^2} = -[44, 4].$$

If in addition we suppose the gravitational field to be a quasi-static field, by confining ourselves to the case where the motion of the matter generating the gravitational field is but slow (in comparison with the velocity of the propagation of light), we may neglect on the right-hand side differentiations with respect to the time in comparison with those with respect to the space co-ordinates, so that we have

$$\frac{d^2x_\tau}{dt^2} = -\tfrac{1}{2}\frac{\partial g_{44}}{\partial x_\tau} \quad (\tau = 1, 2, 3) \quad . \quad . \quad (67)$$

This is the equation of motion of the material point according to Newton's theory, in which $\frac{1}{2}g_{44}$ plays the part of the gravitational potential. What is remarkable in this result is that the component g_{44} of the fundamental tensor alone defines, to a first approximation, the motion of the material point.

We now turn to the field equations (53). Here we have to take into consideration that the energy-tensor of "matter" is almost exclusively defined by the density of matter in the narrower sense, i.e. by the second term of the right-hand side of (58) [or, respectively, (58a) or (58b)]. If we form the approximation in question, all the components vanish with the one exception of $T_{44} = \rho = T$. On the left-hand side of (53) the second term is a small quantity of second order; the first yields, to the approximation in question,

$$\frac{\partial}{\partial x_1}[\mu\nu, 1] + \frac{\partial}{\partial x_2}[\mu\nu, 2] + \frac{\partial}{\partial x_3}[\mu\nu, 3] - \frac{\partial}{\partial x_4}[\mu\nu, 4].$$

For $\mu = \nu = 4$, this gives, with the omission of terms differentiated with respect to time,

$$-\tfrac{1}{2}\left(\frac{\partial^2 g_{44}}{\partial x_1^2} + \frac{\partial^2 g_{44}}{\partial x_2^2} + \frac{\partial^2 g_{44}}{\partial x_3^2}\right) = -\tfrac{1}{2}\nabla^2 g_{44}.$$

The last of equations (53) thus yields

$$\nabla^2 g_{44} = \kappa\rho \quad . \quad . \quad . \quad . \quad (68)$$

The equations (67) and (68) together are equivalent to Newton's law of gravitation.

By (67) and (68) the expression for the gravitational potential becomes

$$-\frac{\kappa}{8\pi}\int\frac{\rho d\tau}{r} \quad . \quad . \quad . \quad . \quad (68a)$$

while Newton's theory, with the unit of time which we have chosen, gives

$$-\frac{K}{c^2}\int\frac{\rho d\tau}{r}$$

in which K denotes the constant $6{\cdot}7 \times 10^{-8}$, usually called the constant of gravitation. By comparison we obtain

$$\kappa = \frac{8\pi K}{c^2} = 1{\cdot}87 \times 10^{-27} \quad . \quad . \quad (69)$$

§ 22. Behaviour of Rods and Clocks in the Static Gravitational Field. Bending of Light-rays. Motion of the Perihelion of a Planetary Orbit

To arrive at Newton's theory as a first approximation we had to calculate only one component, g_{44}, of the ten $g_{\mu\nu}$ of the gravitational field, since this component alone enters into the first approximation, (67), of the equation for the motion of the material point in the gravitational field. From this, however, it is already apparent that other components of the $g_{\mu\nu}$ must differ from the values given in (4) by small quantities of the first order. This is required by the condition $g = -1$.

For a field-producing point mass at the origin of co-ordinates, we obtain, to the first approximation, the radially symmetrical solution

$$\left.\begin{aligned} g_{\rho\sigma} &= -\delta_{\rho\sigma} - \alpha\frac{x_\rho x_\sigma}{r^3} \quad (\rho, \sigma = 1, 2, 3) \\ g_{\rho 4} &= g_{4\rho} = 0 \qquad\qquad (\rho = 1, 2, 3) \\ g_{44} &= 1 - \frac{a}{r} \end{aligned}\right\} \quad . \quad (70)$$

where $\delta_{\rho\sigma}$ is 1 or 0, respectively, accordingly as $\rho = \sigma$ or $\rho \neq \sigma$, and r is the quantity $+\sqrt{x_1^2 + x_2^2 + x_3^2}$. On account of (68a)

$$a = \frac{\kappa M}{4\pi}, \quad . \quad . \quad . \quad . \quad (70a)$$

if M denotes the field-producing mass. It is easy to verify that the field equations (outside the mass) are satisfied to the first order of small quantities.

We now examine the influence exerted by the field of the mass M upon the metrical properties of space. The relation

$$ds^2 = g_{\mu\nu}dx_\mu dx_\nu.$$

always holds between the "locally" (§ 4) measured lengths and times ds on the one hand, and the differences of co-ordinates dx_ν on the other hand.

For a unit-measure of length laid "parallel" to the axis of x, for example, we should have to set $ds^2 = -1$; $dx_2 = dx_3 = dx_4 = 0$. Therefore $-1 = g_{11}dx_1^2$. If, in addition, the unit-measure lies on the axis of x, the first of equations (70) gives

$$g_{11} = -\left(1 + \frac{a}{r}\right).$$

From these two relations it follows that, correct to a first order of small quantities,

$$dx = 1 - \frac{a}{2r} \quad . \quad . \quad . \quad . \quad (71)$$

The unit measuring-rod thus appears a little shortened in relation to the system of co-ordinates by the presence of the gravitational field, if the rod is laid along a radius.

In an analogous manner we obtain the length of co-ordinates in tangential direction if, for example, we set

$ds^2 = -1$; $dx_1 = dx_3 = dx_4 = 0$; $x_1 = r$, $x_2 = x_3 = 0$.

The result is

$$-1 = g_{22}dx_2^2 = -dx_2^2 \quad . \quad . \quad . \quad (71a)$$

With the tangential position, therefore, the gravitational field of the point of mass has no influence on the length of a rod.

Thus Euclidean geometry does not hold even to a first approximation in the gravitational field, if we wish to take one and the same rod, independently of its place and orientation, as a realization of the same interval; although, to be sure, a glance at (70a) and (69) shows that the deviations to be expected are much too slight to be noticeable in measurements of the earth's surface.

Further, let us examine the rate of a unit clock, which is arranged to be at rest in a static gravitational field. Here we have for a clock period $ds = 1$; $dx_1 = dx_2 = dx_3 = 0$

Therefore

$$1 = g_{44}dx_4^2;$$

$$dx_4 = \frac{1}{\sqrt{g_{44}}} = \frac{1}{\sqrt{(1 + (g_{44} - 1))}} = 1 - \tfrac{1}{2}(g_{44} - 1)$$

or

$$dx_4 = 1 + \frac{\kappa}{8\pi}\int \rho\frac{d\tau}{r} \quad . \quad . \quad . \quad (72)$$

Thus the clock goes more slowly if set up in the neighbourhood of ponderable masses. From this it follows that the spectral lines of light reaching us from the surface of large stars must appear displaced towards the red end of the spectrum.*

We now examine the course of light-rays in the static gravitational field. By the special theory of relativity the velocity of light is given by the equation

$$-dx_1^2 - dx_2 - dx_3^2 + dx_4^2 = 0$$

and therefore by the general theory of relativity by the equation

$$ds^2 = g_{\mu\nu}dx_\mu dx_\nu = 0 \quad . \quad . \quad . \quad (73)$$

If the direction, i.e. the ratio $dx_1 : dx_2 : dx_3$ is given, equation (73) gives the quantities

$$\frac{dx_1}{dx_4}, \frac{dx_2}{dx_4}, \frac{dx_3}{dx_4}$$

and accordingly the velocity

$$\sqrt{\left(\frac{dx_1}{dx_4}\right)^2 + \left(\frac{dx_2}{dx_4}\right)^2 + \left(\frac{dx_3}{dx_4}\right)^2} = \gamma$$

defined in the sense of Euclidean geometry. We easily recognize that the course of the light-rays must be bent with regard to the system of co-ordinates, if the $g_{\mu\nu}$ are not constant. If n is a direction perpendicular to the propagation of light, the Huyghens principle shows that the light-ray, envisaged in the plane (γ, n), has the curvature $-\partial\gamma/\partial n$.

We examine the curvature undergone by a ray of light passing by a mass M at the distance Δ. If we choose the system of co-ordinates in agreement with the accompanying diagram, the total bending of the ray (calculated positively if

* According to E. Freundlich, spectroscopical observations on fixed stars of certain types indicate the existence of an effect of this kind, but a crucial test of this consequence has not yet been made.

concave towards the origin) is given in sufficient approximation by

$$B = \int_{-\infty}^{+\infty} \frac{\partial \gamma}{\partial x_1} dx_2,$$

while (73) and (70) give

$$\gamma = \sqrt{\left(-\frac{g_{44}}{g_{22}}\right)} = 1 - \frac{a}{2r}\left(1 + \frac{x_2^2}{r^2}\right).$$

Carrying out the calculation, this gives

$$B = \frac{2a}{\Delta} = \frac{\kappa M}{2\pi\Delta}. \quad . \quad . \quad . \quad (74)$$

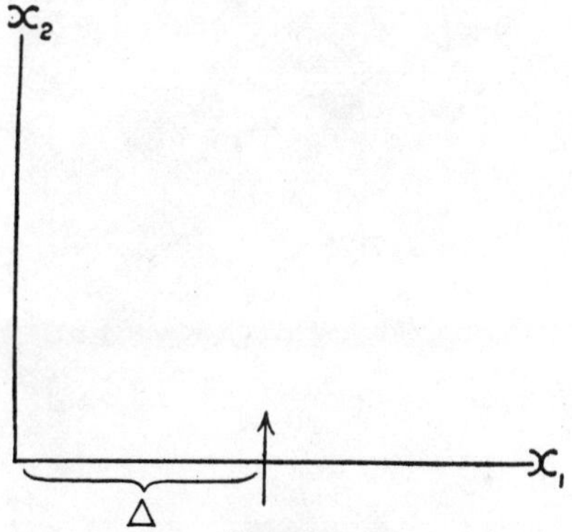

FIG. 8.

According to this, a ray of light going past the sun undergoes a deflexion of 1·7″; and a ray going past the planet Jupiter a deflexion of about ·02″.

If we calculate the gravitational field to a higher degree of approximation, and likewise with corresponding accuracy the orbital motion of a material point of relatively infinitely small mass, we find a deviation of the following kind from the Kepler-Newton laws of planetary motion. The orbital ellipse of a planet undergoes a slow rotation, in the direction of motion, of amount

$$\epsilon = 24\pi^3 \frac{a^2}{T^2c^2(1 - e^2)} \quad . \quad . \quad . \quad (75)$$

per revolution. In this formula a denotes the major semi-axis, c the velocity of light in the usual measurement, e the eccentricity, T the time of revolution in seconds.*

Calculation gives for the planet Mercury a rotation of the orbit of 43″ per century, corresponding exactly to astronomical observation (Leverrier); for the astronomers have discovered in the motion of the perihelion of this planet, after allowing for disturbances by other planets, an inexplicable remainder of this magnitude.

* For the calculation I refer to the original papers: A. Einstein, Sitzungsber. d. Preuss. Akad. d. Wiss., 1915, p. 831; K. Schwarzschild, *ibid.*, 1916, p. 189.

HAMILTON'S PRINCIPLE AND THE GENERAL THEORY OF RELATIVITY

BY

A. EINSTEIN

Translated from "Hamiltonsches Princip und allgemeine Relativitätstheorie," Sitzungsberichte der Preussischen Akad. d. Wissenschaften, 1916.

HAMILTON'S PRINCIPLE AND THE GENERAL THEORY OF RELATIVITY

BY A. EINSTEIN

THE general theory of relativity has recently been given in a particularly clear form by H. A. Lorentz and D. Hilbert,* who have deduced its equations from one single principle of variation. The same thing will be done in the present paper. But my purpose here is to present the fundamental connexions in as perspicuous a manner as possible, and in as general terms as is permissible from the point of view of the general theory of relativity. In particular we shall make as few specializing assumptions as possible, in marked contrast to Hilbert's treatment of the subject. On the other hand, in antithesis to my own most recent treatment of the subject, there is to be complete liberty in the choice of the system of co-ordinates.

§ 1. The Principle of Variation and the Field-equations of Gravitation and Matter

Let the gravitational field be described as usual by the tensor † of the $g_{\mu\nu}$ (or the $g^{\mu\nu}$); and matter, including the electromagnetic field, by any number of space-time functions $q_{(\rho)}$. How these functions may be characterized in the theory of invariants does not concern us. Further, let $\mathfrak{H}$ be a function of the

$$g^{\mu\nu},\ g^{\mu\nu}_{\sigma}\left(=\frac{\partial g^{\mu\nu}}{\partial x_{\sigma}}\right) \text{ and } g^{\mu\nu}_{\sigma\tau}\left(=\frac{\partial^2 g^{\mu\nu}}{\partial x_{\sigma}\partial x_{\tau}}\right), \text{ the } q_{(\rho)} \text{ and } q_{(\rho)\alpha}\left(=\frac{\partial q_{(\rho)}}{\partial x_{\alpha}}\right).$$

* Four papers by Lorentz in the Publications of the Koninkl. Akad. van Wetensch. te Amsterdam, 1915 and 1916; D. Hilbert, Göttinger Nachr., 1915, Part 3.

† No use is made for the present of the tensor character of the $g_{\mu\nu}$.

The principle of variation

$$\delta\int \mathfrak{H} d\tau = 0 \quad . \quad . \quad . \quad . \quad (1)$$

then gives us as many differential equations as there are functions $g_{\mu\nu}$ and $q_{(\rho)}$ to be defined, if the $g^{\mu\nu}$ and $q_{(\rho)}$ are varied independently of one another, and in such a way that at the limits of integration the $\delta q_{(\rho)}$, $\delta g^{\mu\nu}$, and $\frac{\partial}{\partial x_\sigma}(\delta g_{\mu\nu})$ all vanish.

We will now assume that $\mathfrak{H}$ is linear in the $g_{\sigma\tau}$, and that the coefficients of the $g^{\mu\nu}_{\sigma\tau}$ depend only on the $g^{\mu\nu}$. We may then replace the principle of variation (1) by one which is more convenient for us. For by appropriate partial integration we obtain

$$\int \mathfrak{H} d\tau = \int \mathfrak{H}^* d\tau + \mathrm{F} \quad . \quad . \quad . \quad (2)$$

where F denotes an integral over the boundary of the domain in question, and $\mathfrak{H}^*$ depends only on the $g^{\mu\nu}$, $g^{\mu\nu}_{\sigma}$, $q_{(\rho)}$, $q_{(\rho)\alpha}$, and no longer on the $g^{\mu\nu}_{\sigma\tau}$. From (2) we obtain, for such variations as are of interest to us,

$$\delta\int \mathfrak{H} d\tau = \delta\int \mathfrak{H}^* d\tau, \quad . \quad . \quad . \quad (3)$$

so that we may replace our principle of variation (1) by the more convenient form

$$\delta\int \mathfrak{H}^* d\tau = 0. \quad . \quad . \quad . \quad (1a)$$

By carrying out the variation of the $g^{\mu\nu}$ and the $q_{(\rho)}$ we obtain, as field-equations of gravitation and matter, the equations †

$$\frac{\partial}{\partial x_\alpha}\left(\frac{\partial \mathfrak{H}^*}{\partial g^{\mu\nu}_{\alpha}}\right) - \frac{\partial \mathfrak{H}^*}{\partial g^{\mu\nu}} = 0 \quad . \quad . \quad . \quad (4)$$

$$\frac{\partial}{\partial x_\alpha}\left(\frac{\partial \mathfrak{H}^*}{\partial q_{(\rho)\alpha}}\right) - \frac{\partial \mathfrak{H}^*}{\partial q_{(\rho)}} = 0 \quad . \quad . \quad . \quad (5)$$

† For brevity the summation symbols are omitted in the formulæ. Indices occurring twice in a term are always to be taken as summed. Thus in (4), for example, $\frac{\partial}{\partial x_\alpha}\left(\frac{\partial \mathfrak{H}^*}{\partial g^{\mu\nu}_{\alpha}}\right)$ denotes the term $\sum_\alpha \frac{\partial}{\partial x_\alpha}\left(\frac{\partial \mathfrak{H}^*}{\partial g^{\mu\nu}_{\alpha}}\right)$.

§ 2. Separate Existence of the Gravitational Field

If we make no restrictive assumption as to the manner in which $\mathfrak{H}$ depends on the $g^{\mu\nu}$, $g^{\mu\nu}_{\sigma}$, $g^{\mu\nu}_{\sigma\tau}$, $q_{(\rho)}$, $q_{(\rho)\alpha}$, the energy-components cannot be divided into two parts, one belonging to the gravitational field, the other to matter. To ensure this feature of the theory, we make the following assumption

$$\mathfrak{H} = \mathfrak{G} + \mathfrak{M} \quad . \quad . \quad . \quad . \quad (6)$$

where $\mathfrak{G}$ is to depend only on the $g^{\mu\nu}$, $g^{\mu\nu}_{\sigma}$, $g^{\mu\nu}_{\sigma\tau}$, and $\mathfrak{M}$ only on $g^{\mu\nu}$, $q_{(\rho)}$, $q_{(\rho)\alpha}$. Equations (4), (4a) then assume the form

$$\frac{\partial}{\partial x_\alpha}\left(\frac{\partial \mathfrak{G}^*}{\partial g^{\mu\nu}_\alpha}\right) - \frac{\partial \mathfrak{G}^*}{\partial g^{\mu\nu}} = \frac{\partial \mathfrak{M}}{\partial g^{\mu\nu}} \quad . \quad . \quad . \quad (7)$$

$$\frac{\partial}{\partial x_\alpha}\left(\frac{\partial \mathfrak{M}}{\partial q_{(\rho)\alpha}}\right) - \frac{\partial \mathfrak{M}}{\partial q_{(\rho)}} = 0 \quad . \quad . \quad . \quad (8)$$

Here $\mathfrak{G}^*$ stands in the same relation to $\mathfrak{G}$ as $\mathfrak{H}^*$ to $\mathfrak{H}$.

It is to be noted carefully that equations (8) or (5) would have to give way to others, if we were to assume $\mathfrak{M}$ or $\mathfrak{H}$ to be also dependent on derivatives of the $q_{(\rho)}$ of order higher than the first. Likewise it might be imaginable that the $q_{(\rho)}$ would have to be taken, not as independent of one another, but as connected by conditional equations. All this is of no importance for the following developments, as these are based solely on the equations (7), which have been found by varying our integral with respect to the $g^{\mu\nu}$.

§ 3. Properties of the Field Equations of Gravitation Conditioned by the Theory of Invariants

We now introduce the assumption that

$$ds^2 = g_{\mu\nu}dx_\mu dx_\nu \quad . \quad . \quad . \quad . \quad (9)$$

is an invariant. This determines the transformational character of the $g_{\mu\nu}$. As to the transformational character of the $q_{(\rho)}$, which describe matter, we make no supposition. On the other hand, let the functions $\mathrm{H} = \frac{\mathfrak{H}}{\sqrt{-g}}$, as well as

$G = \frac{\mathfrak{G}}{\sqrt{-g}}$, and $M = \frac{\mathfrak{M}}{\sqrt{-g}}$, be invariants in relation to any substitutions of space-time co-ordinates. From these assumptions follows the general covariance of the equations (7) and (8), deduced from (1). It further follows that G (apart from a constant factor) must be equal to the scalar of Riemann's tensor of curvature; because there is no other invariant with the properties required for G.† Thereby $\mathfrak{G}^*$ is also perfectly determined, and consequently the left-hand side of field equation (7) as well.‡

From the general postulate of relativity there follow certain properties of the function $\mathfrak{G}^*$ which we shall now deduce. For this purpose we carry through an infinitesimal transformation of the co-ordinates, by setting

$$x'_\nu = x_\nu + \Delta x_\nu \quad . \quad . \quad . \quad . \quad (10)$$

where the Δx_ν are arbitrary, infinitely small functions of the co-ordinates, and x'_ν are the co-ordinates, in the new system, of the world-point having the co-ordinates x_ν in the original system. As for the co-ordinates, so too for any other magnitude ψ, a law of transformation holds good, of the type

$$\psi' = \psi + \Delta\psi,$$

where $\Delta\psi$ must always be expressible by the Δx_ν. From the covariant property of the $g^{\mu\nu}$ we easily deduce for the $g^{\mu\nu}$ and $g^{\mu\nu}_\sigma$ the laws of transformation

$$\Delta g^{\mu\nu} = g^{\mu\alpha}\frac{\partial(\Delta x_\nu)}{\partial x_\alpha} + g^{\nu\alpha}\frac{\partial(\Delta x_\mu)}{\partial x_\alpha} \quad . \quad . \quad (11)$$

$$\Delta g^{\mu\nu}_\sigma = \frac{\partial(\Delta g^{\mu\nu})}{\partial x_\sigma} - g^{\mu\nu}_\alpha\frac{\partial(\Delta x_\alpha)}{\partial x_\sigma} \quad . \quad . \quad . \quad (12)$$

Since $\mathfrak{G}^*$ depends only on the $g^{\mu\nu}$ and $g^{\mu\nu}_\sigma$, it is possible, with the help of (11) and (12), to calculate $\Delta\mathfrak{G}^*$. We thus obtain the equation

$$\sqrt{-g}\Delta\left(\frac{\mathfrak{G}^*}{\sqrt{-g}}\right) = S^\nu_\sigma\frac{\partial(\Delta x_\sigma)}{\partial x_\nu} + 2\frac{\partial\mathfrak{G}^*}{\partial g^{\mu\sigma}_\alpha}g^{\mu\nu}\frac{\partial^2\Delta x_\sigma}{\partial x_\nu\partial x_\alpha}, \quad (13)$$

† Herein is to be found the reason why the general postulate of relativity leads to a very definite theory of gravitation.

‡ By performing partial integration we obtain

$$\mathfrak{G}^* = \sqrt{-g}\, g^{\mu\nu}[\{\mu\alpha, \beta\}\{\nu\beta, \alpha\} - \{\mu\nu, \alpha\}\{\alpha\beta, \beta\}].$$

where for brevity we have set

$$S^\nu_\sigma = 2\frac{\partial \mathfrak{G}^*}{\partial g^{\mu\sigma}}g^{\mu\nu} + 2\frac{\partial \mathfrak{G}^*}{\partial g^{\mu\sigma}_\alpha}g^{\mu\nu}_\alpha + \mathfrak{G}^*\delta^\nu_\sigma - \frac{\partial \mathfrak{G}^*}{\partial g^{\mu\alpha}_\nu}g^{\mu\alpha}_\sigma. \qquad (14)$$

From these two equations we draw two inferences which are important for what follows. We know that $\frac{\mathfrak{G}}{\sqrt{-g}}$ is an invariant with respect to any substitution, but we do not know this of $\frac{\mathfrak{G}^*}{\sqrt{-g}}$. It is easy to demonstrate, however, that the latter quantity is an invariant with respect to any *linear* substitutions of the co-ordinates. Hence it follows that the right side of (13) must always vanish if all $\frac{\partial^2 \Delta x_\sigma}{\partial x_\nu \partial x_\alpha}$ vanish. Consequently $\mathfrak{G}^*$ must satisfy the identity

$$S^\nu_\sigma \equiv 0 \quad . \quad . \quad . \quad . \quad (15)$$

If, further, we choose the Δx_ν so that they differ from zero only in the interior of a given domain, but in infinitesimal proximity to the boundary they vanish, then, with the transformation in question, the value of the boundary integral occurring in equation (2) does not change. Therefore $\Delta F = 0$, and, in consequence,†

$$\Delta\int \mathfrak{G} d\tau = \Delta\int \mathfrak{G}^* d\tau.$$

But the left-hand side of the equation must vanish, since both $\frac{\mathfrak{G}}{\sqrt{-g}}$ and $\sqrt{-g}\, d\tau$ are invariants. Consequently the right-hand side also vanishes. Thus, taking (14), (15), and (16) into consideration, we obtain, in the first place, the equation

$$\int \frac{\partial \mathfrak{G}^*}{\partial g^{\mu\sigma}_\alpha} g^{\mu\nu} \frac{\partial^2(\Delta x_\sigma)}{\partial x_\nu \partial x_\alpha} d\tau = 0 \quad . \quad . \quad . \quad (16)$$

Transforming this equation by two partial integrations, and having regard to the liberty of choice of the Δx_σ, we obtain

† By the introduction of the quantities $\mathfrak{G}$ and $\mathfrak{G}^*$ instead of $\mathfrak{H}$ and $\mathfrak{H}^*$.

the identity

$$\frac{\partial^2}{\partial x_\nu \partial x_\alpha}\left(g^{\mu\nu}\frac{\partial \mathfrak{G}^*}{\partial g^{\mu\sigma}_\alpha}\right) \equiv 0 \quad . \quad . \quad . \quad (17)$$

From the two identities (16) and (17), which result from the invariance of $\frac{\mathfrak{G}}{\sqrt{-g}}$, and therefore from the postulate of general relativity, we now have to draw conclusions.

We first transform the field equations (7) of gravitation by mixed multiplication by $g^{\mu\sigma}$. We then obtain (by interchanging the indices σ and ν), as equivalents of the field equations (7), the equations

$$\frac{\partial}{\partial x_\alpha}\left(g^{\mu\nu}\frac{\partial \mathfrak{G}^*}{\partial g^{\mu\sigma}_\alpha}\right) = -(\mathfrak{T}^\nu_\sigma + \mathfrak{t}^\nu_\sigma) \quad . \quad . \quad (18)$$

where we have set

$$\mathfrak{T}^\nu_\sigma = -\frac{\partial \mathfrak{M}}{\partial g^{\mu\sigma}}g^{\mu\nu} \quad . \quad . \quad . \quad . \quad . \quad . \quad . \quad (19)$$

$$\mathfrak{t}^\nu_\sigma = -\left(\frac{\partial \mathfrak{G}^*}{\partial g^{\mu\sigma}_\alpha}g^{\mu\nu}_\alpha + \frac{\partial \mathfrak{G}^*}{\partial g^{\mu\sigma}}g^{\mu\nu}\right) = \tfrac{1}{2}\left(\mathfrak{G}^*\delta^\nu_\sigma - \frac{\partial \mathfrak{G}^*}{\partial g^{\mu\alpha}_\nu}g^{\mu\alpha}_\sigma\right) \quad (20)$$

The last expression for $\mathfrak{t}^\nu_\sigma$ is vindicated by (14) and (15). By differentiation of (18) with respect to x_ν, and summation for ν, there follows, in view of (17),

$$\frac{\partial}{\partial x_\nu}(\mathfrak{T}^\nu_\sigma + \mathfrak{t}^\nu_\sigma) = 0 \quad . \quad . \quad . \quad (21)$$

Equation (21) expresses the conservation of momentum and energy. We call $\mathfrak{T}^\nu_\sigma$ the components of the energy of matter, $\mathfrak{t}^\nu_\sigma$ the components of the energy of the gravitational field.

Having regard to (20), there follows from the field equations (7) of gravitation, by multiplication by $g^{\mu\nu}_\sigma$ and summation with respect to μ and ν,

$$\frac{\partial \mathfrak{t}^\nu_\sigma}{\partial x_\nu} + \tfrac{1}{2}g^{\mu\nu}_\sigma\frac{\partial \mathfrak{M}}{\partial g^{\mu\nu}} = 0,$$

or, in view of (19) and (21),

$$\frac{\partial \mathfrak{T}_\sigma^\nu}{\partial x_\nu} + \tfrac{1}{2} g_\sigma^{\mu\nu} \mathfrak{T}_{\mu\nu} = 0 \quad . \quad . \quad . \quad (22)$$

where $\mathfrak{T}_{\mu\nu}$ denotes the quantities $g_{\nu\sigma}\mathfrak{T}_\mu^\sigma$. These are four equations which the energy-components of matter have to satisfy.

It is to be emphasized that the (generally covariant) laws of conservation (21) and (22) are deduced from the field equations (7) of gravitation, in combination with the postulate of general covariance (relativity) *alone*, without using the field equations (8) for material phenomena.

COSMOLOGICAL CONSIDERATIONS ON THE GENERAL THEORY OF RELATIVITY

BY

A. EINSTEIN

Translated from "Kosmologische Betrachtungen zur allgemeinen Relativitätstheorie," Sitzungsberichte der Preussischen Akad. d. Wissenschaften, 1917.

COSMOLOGICAL CONSIDERATIONS ON THE GENERAL THEORY OF RELATIVITY

By A. EINSTEIN

It is well known that Poisson's equation

$$\nabla^2\phi = 4\pi K\rho \quad . \quad . \quad . \quad . \quad (1)$$

in combination with the equations of motion of a material point is not as yet a perfect substitute for Newton's theory of action at a distance. There is still to be taken into account the condition that at spatial infinity the potential ϕ tends toward a fixed limiting value. There is an analogous state of things in the theory of gravitation in general relativity. Here, too, we must supplement the differential equations by limiting conditions at spatial infinity, if we really have to regard the universe as being of infinite spatial extent.

In my treatment of the planetary problem I chose these limiting conditions in the form of the following assumption: it is possible to select a system of reference so that at spatial infinity all the gravitational potentials $g_{\mu\nu}$ become constant. But it is by no means evident *a priori* that we may lay down the same limiting conditions when we wish to take larger portions of the physical universe into consideration. In the following pages the reflexions will be given which, up to the present, I have made on this fundamentally important question.

§ 1. The Newtonian Theory

It is well known that Newton's limiting condition of the constant limit for ϕ at spatial infinity leads to the view that the density of matter becomes zero at infinity. For we imagine that there may be a place in universal space round about which the gravitational field of matter, viewed on a large scale, possesses spherical symmetry. It then follows from Poisson's equation that, in order that ϕ may tend to a limit at infinity, the mean density ρ must decrease toward zero more rapidly than $1/r^2$ as the distance r from the centre increases.* In this sense, therefore, the universe according to Newton is finite, although it may possess an infinitely great total mass.

From this it follows in the first place that the radiation emitted by the heavenly bodies will, in part, leave the Newtonian system of the universe, passing radially outwards, to become ineffective and lost in the infinite. May not entire heavenly bodies fare likewise? It is hardly possible to give a negative answer to this question. For it follows from the assumption of a finite limit for ϕ at spatial infinity that a heavenly body with finite kinetic energy is able to reach spatial infinity by overcoming the Newtonian forces of attraction. By statistical mechanics this case must occur from time to time, as long as the total energy of the stellar system—transferred to one single star—is great enough to send that star on its journey to infinity, whence it never can return.

We might try to avoid this peculiar difficulty by assuming a very high value for the limiting potential at infinity. That would be a possible way, if the value of the gravitational potential were not itself necessarily conditioned by the heavenly bodies. The truth is that we are compelled to regard the occurrence of any great differences of potential of the gravitational field as contradicting the facts. These differences must really be of so low an order of magnitude that the stellar velocities generated by them do not exceed the velocities actually observed.

If we apply Boltzmann's law of distribution for gas molecules to the stars, by comparing the stellar system with a gas in thermal equilibrium, we find that the Newtonian stellar system cannot exist at all. For there is a finite ratio of densities corresponding to the finite difference of potential between the centre and spatial infinity. A vanishing of the density at infinity thus implies a vanishing of the density at the centre.

* ρ is the mean density of matter, calculated for a region which is large as compared with the distance between neighbouring fixed stars, but small in comparison with the dimensions of the whole stellar system.

It seems hardly possible to surmount these difficulties on the basis of the Newtonian theory. We may ask ourselves the question whether they can be removed by a modification of the Newtonian theory. First of all we will indicate a method which does not in itself claim to be taken seriously; it merely serves as a foil for what is to follow. In place of Poisson's equation we write

$$\nabla^2\phi - \lambda\phi = 4\pi\kappa\rho \quad . \quad . \quad . \quad (2)$$

where λ denotes a universal constant. If ρ_0 be the uniform density of a distribution of mass, then

$$\phi = -\frac{4\pi\kappa}{\lambda}\rho_0 \quad . \quad . \quad . \quad . \quad (3)$$

is a solution of equation (2). This solution would correspond to the case in which the matter of the fixed stars was distributed uniformly through space, if the density ρ_0 is equal to the actual mean density of the matter in the universe. The solution then corresponds to an infinite extension of the central space, filled uniformly with matter. If, without making any change in the mean density, we imagine matter to be non-uniformly distributed locally, there will be, over and above the ϕ with the constant value of equation (3), an additional ϕ, which in the neighbourhood of denser masses will so much the more resemble the Newtonian field as $\lambda\phi$ is smaller in comparison with $4\pi\kappa\rho$.

A universe so constituted would have, with respect to its gravitational field, no centre. A decrease of density in spatial infinity would not have to be assumed, but both the mean potential and mean density would remain constant to infinity. The conflict with statistical mechanics which we found in the case of the Newtonian theory is not repeated. With a definite but extremely small density, matter is in equilibrium, without any internal material forces (pressures) being required to maintain equilibrium.

§ 2. The Boundary Conditions According to the General Theory of Relativity

In the present paragraph I shall conduct the reader over the road that I have myself travelled, rather a rough and winding road, because otherwise I cannot hope that he will take much interest in the result at the end of the journey. The conclusion I shall arrive at is that the field equations of gravitation which I have championed hitherto still need a slight modification, so that on the basis of the general theory of relativity those fundamental difficulties may be avoided which have been set forth in § 1 as confronting the Newtonian theory. This modification corresponds perfectly to the transition from Poisson's equation (1) to equation (2) of § 1. We finally infer that boundary conditions in spatial infinity fall away altogether, because the universal continuum in respect of its spatial dimensions is to be viewed as a self-contained continuum of finite spatial (three-dimensional) volume.

The opinion which I entertained until recently, as to the limiting conditions to be laid down in spatial infinity, took its stand on the following considerations. In a consistent theory of relativity there can be no inertia *relatively to "space,"* but only an inertia of masses *relatively to one another*. If, therefore, I have a mass at a sufficient distance from all other masses in the universe, its inertia must fall to zero. We will try to formulate this condition mathematically.

According to the general theory of relativity the negative momentum is given by the first three components, the energy by the last component of the covariant tensor multiplied by $\sqrt{-g}$

$$m\sqrt{-g}\;\; g_{\mu\alpha}\frac{dx_\alpha}{ds} \quad . \quad . \quad . \quad . \quad (4)$$

where, as always, we set

$$ds^2 = g_{\mu\nu}dx_\mu dx_\nu \quad . \quad . \quad . \quad . \quad (5)$$

In the particularly perspicuous case of the possibility of choosing the system of co-ordinates so that the gravitational field at every point is spatially isotropic, we have more simply

$$ds^2 = -\mathrm{A}(dx_1^2 + dx_2^2 + dx_3^2) + \mathrm{B}dx_4^2.$$

If, moreover, at the same time

$$\sqrt{-g} = 1 = \sqrt{\mathrm{A}^3\mathrm{B}}$$

we obtain from (4), to a first approximation for small velocities,

$$m\frac{\mathrm{A}}{\sqrt{\mathrm{B}}}\frac{dx_1}{dx_4},\; m\frac{\mathrm{A}}{\sqrt{\mathrm{B}}}\frac{dx_2}{dx_4},\; m\frac{\mathrm{A}}{\sqrt{\mathrm{B}}}\frac{dx_3}{dx_4}$$

for the components of momentum, and for the energy (in the static case)

$$m\sqrt{B}.$$

From the expressions for the momentum, it follows that $m\frac{A}{\sqrt{B}}$ plays the part of the rest mass. As m is a constant peculiar to the point of mass, independently of its position, this expression, if we retain the condition $\sqrt{g -} = 1$ at spatial infinity, can vanish only when A diminishes to zero, while B increases to infinity. It seems, therefore, that such a degeneration of the co-efficients $g_{\mu\nu}$ is required by the postulate of relativity of all inertia. This requirement implies that the potential energy $m\sqrt{B}$ becomes infinitely great at infinity. Thus a point of mass can never leave the system; and a more detailed investigation shows that the same thing applies to light-rays. A system of the universe with such behaviour of the gravitational potentials at infinity would not therefore run the risk of wasting away which was mooted just now in connexion with the Newtonian theory.

I wish to point out that the simplifying assumptions as to the gravitational potentials on which this reasoning is based, have been introduced merely for the sake of lucidity. It is possible to find general formulations for the behaviour of the $g_{\mu\nu}$ at infinity which express the essentials of the question without further restrictive assumptions.

At this stage, with the kind assistance of the mathematician J. Grommer, I investigated centrally symmetrical, static gravitational fields, degenerating at infinity in the way mentioned. The gravitational potentials $g_{\mu\nu}$ were applied, and from them the energy-tensor $T_{\mu\nu}$ of matter was calculated on the basis of the field equations of gravitation. But here it proved that for the system of the fixed stars no boundary conditions of the kind can come into question at all, as was also rightly emphasized by the astronomer de Sitter recently.

For the contravariant energy-tensor $T^{\mu\nu}$ of ponderable matter is given by

$$T^{\mu\nu} = \rho\frac{dx_\mu}{ds}\frac{dx_\nu}{ds},$$

where ρ is the density of matter in natural measure. With an appropriate choice of the system of co-ordinates the stellar velocities are very small in comparison with that of light. We may, therefore, substitute $\sqrt{g_{44}}\,dx_4$ for ds. This shows us that all components of $T^{\mu\nu}$ must be very small in comparison with the last component T^{44}. But it was quite impossible to reconcile this condition with the chosen boundary conditions. In the retrospect this result does not appear astonishing. The fact of the small velocities of the stars allows the conclusion that wherever there are fixed stars, the gravitational potential (in our case $\sqrt{B}$) can never be much greater than here on earth. This follows from statistical reasoning, exactly as in the case of the Newtonian theory. At any rate, our calculations have convinced me that such conditions of degeneration for the $g_{\mu\nu}$ in spatial infinity may not be postulated.

After the failure of this attempt, two possibilities next present themselves.

(*a*) We may require, as in the problem of the planets, that, with a suitable choice of the system of reference, the $g_{\mu\nu}$ in spatial infinity approximate to the values

$$\begin{matrix} -1 & 0 & 0 & 0 \\ 0 & -1 & 0 & 0 \\ 0 & 0 & -1 & 0 \\ 0 & 0 & 0 & 1 \end{matrix}$$

(*b*) We may refrain entirely from laying down boundary conditions for spatial infinity claiming general validity; but at the spatial limit of the domain under consideration we have to give the $g_{\mu\nu}$ separately in each individual case, as hitherto we were accustomed to give the initial conditions for time separately.

The possibility (*b*) holds out no hope of solving the problem, but amounts to giving it up. This is an incontestable position, which is taken up at the present time by de Sitter.* But I must confess that such a complete resignation in this fundamental question is for me a difficult thing. I should not make up my mind to it until every effort to make headway toward a satisfactory view had proved to be vain.

Possibility (*a*) is unsatisfactory in more respects than one.

* de Sitter, Akad. van Wetensch. te Amsterdam, 8 Nov., 1916.

In the first place those boundary conditions pre-suppose a definite choice of the system of reference, which is contrary to the spirit of the relativity principle. Secondly, if we adopt this view, we fail to comply with the requirement of the relativity of inertia. For the inertia of a material point of mass m (in natural measure) depends upon the $g_{\mu\nu}$; but these differ but little from their postulated values, as given above, for spatial infinity. Thus inertia would indeed be *influenced*, but would not be *conditioned* by matter (present in finite space). If only one single point of mass were present, according to this view, it would possess inertia, and in fact an inertia almost as great as when it is surrounded by the other masses of the actual universe. Finally, those statistical objections must be raised against this view which were mentioned in respect of the Newtonian theory.

From what has now been said it will be seen that I have not succeeded in formulating boundary conditions for spatial infinity. Nevertheless, there is still a possible way out, without resigning as suggested under (*b*). For if it were possible to regard the universe as a continuum which is *finite* (*closed*) *with respect to its spatial dimensions*, we should have no need at all of any such boundary conditions. We shall proceed to show that both the general postulate of relativity and the fact of the small stellar velocities are compatible with the hypothesis of a spatially finite universe; though certainly, in order to carry through this idea, we need a generalizing modification of the field equations of gravitation.

§ 3. The Spatially Finite Universe with a Uniform Distribution of Matter

According to the general theory of relativity the metrical character (curvature) of the four-dimensional space-time continuum is defined at every point by the matter at that point and the state of that matter. Therefore, on account of the lack of uniformity in the distribution of matter, the metrical structure of this continuum must necessarily be extremely complicated. But if we are concerned with the structure only on a large scale, we may represent matter to ourselves as being uniformly distributed over enormous spaces, so that its density of distribution is a variable function which varies extremely slowly. Thus our procedure will somewhat resemble that of the geodesists who, by means of an ellipsoid, approximate to the shape of the earth's surface, which on a small scale is extremely complicated.

The most important fact that we draw from experience as to the distribution of matter is that the relative velocities of the stars are very small as compared with the velocity of light. So I think that for the present we may base our reasoning upon the following approximative assumption. There is a system of reference relatively to which matter may be looked upon as being permanently at rest. With respect to this system, therefore, the contravariant energy-tensor $T^{\mu\nu}$ of matter is, by reason of (5), of the simple form

$$\left.\begin{matrix} 0 & 0 & 0 & 0 \\ 0 & 0 & 0 & 0 \\ 0 & 0 & 0 & 0 \\ 0 & 0 & 0 & \rho \end{matrix}\right\} \quad . \quad . \quad . \quad (6)$$

The scalar ρ of the (mean) density of distribution may be *a priori* a function of the space co-ordinates. But if we assume the universe to be spatially finite, we are prompted to the hypothesis that ρ is to be independent of locality. On this hypothesis we base the following considerations.

As concerns the gravitational field, it follows from the equation of motion of the material point

$$\frac{d^2x_\nu}{ds^2} + \{\alpha\beta, \nu\}\frac{dx_\alpha}{ds}\frac{dx_\beta}{ds} = 0$$

that a material point in a static gravitational field can remain at rest only when g_{44} is independent of locality. Since, further, we presuppose independence of the time co-ordinate x_4 for all magnitudes, we may demand for the required solution that, for all x_ν,

$$g_{44} = 1 \quad . \quad . \quad . \quad . \quad (7)$$

Further, as always with static problems, we shall have to set

$$g_{14} = g_{24} = g_{34} = 0 \quad . \quad . \quad . \quad (8)$$

It remains now to determine those components of the gravitational potential which define the purely spatial-geometrical relations of our continuum ($g_{11}, g_{12}, \ldots g_{33}$). From our assumption as to the uniformity of distribution of the masses generating the field, it follows that the curvature of the required space must be constant. With this distribution of mass, therefore, the required finite continuum of the x_1, x_2, x_3, with constant x_4, will be a spherical space.

We arrive at such a space, for example, in the following way. We start from a Euclidean space of four dimensions, ξ_1, ξ_2, ξ_3 ξ_4, with a linear element $d\sigma$; let, therefore,

$$d\sigma^2 = d\xi_1^2 + d\xi_2^2 + d\xi_3^2 + d\xi_4^2 \quad . \quad . \quad . \quad (9)$$

In this space we consider the hyper-surface

$$R^2 = \xi_1^2 + \xi_2^2 + \xi_3^2 + \xi_4^2, \quad . \quad . \quad . \quad (10)$$

where R denotes a constant. The points of this hyper-surface form a three-dimensional continuum, a spherical space of radius of curvature R.

The four-dimensional Euclidean space with which we started serves only for a convenient definition of our hyper-surface. Only those points of the hyper-surface are of interest to us which have metrical properties in agreement with those of physical space with a uniform distribution of matter. For the description of this three-dimensional continuum we may employ the co-ordinates ξ_1, ξ_2, ξ_3 (the projection upon the hyper-plane $\xi_4 = 0$) since, by reason of (10), ξ_4 can be expressed in terms of ξ_1, ξ_2, ξ_3. Eliminating ξ_4 from (9), we obtain for the linear element of the spherical space the expression

$$\left.\begin{aligned} d\sigma^2 &= \gamma_{\mu\nu} d\xi_\mu d\xi_\nu \\ \gamma_{\mu\nu} &= \delta_{\mu\nu} + \frac{\xi_\mu \xi_\nu}{R^2 - \rho^2} \end{aligned}\right\} \quad . \quad . \quad . \quad (11)$$

where $\delta_{\mu\nu} = 1$, if $\mu = \nu$; $\delta_{\mu\nu} = 0$, if $\mu \neq \nu$, and $\rho^2 = \xi_1^2 + \xi_2^2 + \xi_3^2$. The co-ordinates chosen are convenient when it is a question of examining the environment of one of the two points $\xi_1 = \xi_2 = \xi_3 = 0$.

Now the linear element of the required four-dimensional space-time universe is also given us. For the potential $g_{\mu\nu}$, both indices of which differ from 4, we have to set

$$g_{\mu\nu} = -\left(\delta_{\mu\nu} + \frac{x_\mu x_\nu}{R^2 - (x_1^2 + x_2^2 + x_3^2)}\right) \quad . \quad (12)$$

which equation, in combination with (7) and (8), perfectly defines the behaviour of measuring-rods, clocks, and light-rays.

§ 4. On an Additional Term for the Field Equations of Gravitation

My proposed field equations of gravitation for any chosen system of co-ordinates run as follows:—

$$\left.\begin{aligned} G_{\mu\nu} &= -\kappa(T_{\mu\nu} - \tfrac{1}{2}g_{\mu\nu}T), \\ G_{\mu\nu} &= -\frac{\partial}{\partial x_\alpha}\{\mu\nu,\alpha\} + \{\mu\alpha,\beta\}\{\nu\beta,\alpha\} \\ &\qquad + \frac{\partial^2 \log\sqrt{-g}}{\partial x_\mu \partial x_\nu} - \{\mu\nu,\alpha\}\frac{\partial \log\sqrt{-g}}{\partial x_\alpha} \end{aligned}\right\} \quad (13)$$

The system of equations (13) is by no means satisfied when we insert for the $g_{\mu\nu}$ the values given in (7), (8), and (12), and for the (contravariant) energy-tensor of matter the values indicated in (6). It will be shown in the next paragraph how this calculation may conveniently be made. So that, if it were certain that the field equations (13) which I have hitherto employed were the only ones compatible with the postulate of general relativity, we should probably have to conclude that the theory of relativity does not admit the hypothesis of a spatially finite universe.

However, the system of equations (14) allows a readily suggested extension which is compatible with the relativity postulate, and is perfectly analogous to the extension of Poisson's equation given by equation (2). For on the left-hand side of field equation (13) we may add the fundamental tensor $g_{\mu\nu}$, multiplied by a universal constant, $-\lambda$, at present unknown, without destroying the general covariance. In place of field equation (13) we write

$$G_{\mu\nu} - \lambda g_{\mu\nu} = -\kappa(T_{\mu\nu} - \tfrac{1}{2}g_{\mu\nu}T) \quad . \quad . \quad (13a)$$

This field equation, with λ sufficiently small, is in any case also compatible with the facts of experience derived from the solar system. It also satisfies laws of conservation of momentum and energy, because we arrive at (13a) in place of (13) by introducing into Hamilton's principle, instead of the scalar of Riemann's tensor, this scalar increased by a universal constant; and Hamilton's principle, of course, guarantees the validity of laws of conservation. It will be shown in § 5 that field equation (13a) is compatible with our conjectures on field and matter.

§ 5. Calculation and Result

Since all points of our continuum are on an equal footing, it is sufficient to carry through the calculation for *one* point, e.g. for one of the two points with the co-ordinates

$$x_1 = x_2 = x_3 = x_4 = 0.$$

Then for the $g_{\mu\nu}$ in (13a) we have to insert the values

$$\begin{matrix} -1 & 0 & 0 & 0 \\ 0 & -1 & 0 & 0 \\ 0 & 0 & -1 & 0 \\ 0 & 0 & 0 & 1 \end{matrix}$$

wherever they appear differentiated only once or not at all. We thus obtain in the first place

$$G_{\mu\nu} = \frac{\partial}{\partial x_1}[\mu\nu, 1] + \frac{\partial}{\partial x_2}[\mu\nu, 2] + \frac{\partial}{\partial x_3}[\mu\nu, 3] + \frac{\partial^2 \log \sqrt{-g}}{\partial x_\mu \partial x_\nu}.$$

From this we readily discover, taking (7), (8), and (13) into account, that all equations (13*a*) are satisfied if the two relations

$$-\frac{2}{R^2} + \lambda = -\frac{\kappa\rho}{2}, \quad -\lambda = -\frac{\kappa\rho}{2},$$

or

$$\lambda = \frac{\kappa\rho}{2} = \frac{1}{R^2} \quad . \quad . \quad . \quad . \quad (14)$$

are fulfilled.

Thus the newly introduced universal constant λ defines both the mean density of distribution ρ which can remain in equilibrium and also the radius R and the volume $2\pi^2 R^3$ of spherical space. The total mass M of the universe, according to our view, is finite, and is in fact

$$M = \rho \, . \, 2\pi^2 R^3 = 4\pi^2 \frac{R}{\kappa} = \pi^2 \sqrt{\frac{32}{\kappa^3 \rho}} \quad . \quad . \quad (15)$$

Thus the theoretical view of the actual universe, if it is in correspondence with our reasoning, is the following. The curvature of space is variable in time and place, according to the distribution of matter, but we may roughly approximate to it by means of a spherical space. At any rate, this view is logically consistent, and from the standpoint of the general theory of relativity lies nearest at hand; whether, from the standpoint of present astronomical knowledge, it is tenable, will not here be discussed. In order to arrive at this consistent view, we admittedly had to introduce an extension of the field equations of gravitation which is not justified by our actual knowledge of gravitation. It is to be emphasized, however, that a positive curvature of space is given by our results, even if the supplementary term is not introduced. That term is necessary only for the purpose of making possible a quasi-static distribution of matter, as required by the fact of the small velocities of the stars.

DO GRAVITATIONAL FIELDS PLAY AN ESSENTIAL PART IN THE STRUCTURE OF THE ELEMENTARY PARTICLES OF MATTER?

BY

A. EINSTEIN

Translated from "Spielen Gravitationsfelder im Aufber der materiellen Elementarteilchen eine wesentliche Rolle?" Sitzungsberichte der Preussischen Akad. d. Wissenschaften, 1919.

DO GRAVITATIONAL FIELDS PLAY AN ESSENTIAL PART IN THE STRUCTURE OF THE ELEMENTARY PARTICLES OF MATTER?

By A. EINSTEIN

NEITHER the Newtonian nor the relativistic theory of gravitation has so far led to any advance in the theory of the constitution of matter. In view of this fact it will be shown in the following pages that there are reasons for thinking that the elementary formations which go to make up the atom are held together by gravitational forces.

§ 1. Defects of the Present View

Great pains have been taken to elaborate a theory which will account for the equilibrium of the electricity constituting the electron. G. Mie, in particular, has devoted deep researches to this question. His theory, which has found considerable support among theoretical physicists, is based mainly on the introduction into the energy-tensor of supplementary terms depending on the components of the electro-dynamic potential, in addition to the energy terms of the Maxwell-Lorentz theory. These new terms, which in outside space are unimportant, are nevertheless effective in the interior of the electrons in maintaining equilibrium against the electric forces of repulsion. In spite of the beauty of the formal structure of this theory, as erected by Mie, Hilbert, and Weyl, its physical results have hitherto been unsatisfactory. On the one hand the multiplicity of possibilities is discouraging, and on the other hand those additional terms have not as yet allowed themselves to be framed in such a simple form that the solution could be satisfactory.

So far the general theory of relativity has made no change in this state of the question. If we for the moment disregard the additional cosmological term, the field equations take the form

$$G_{\mu\nu} - \tfrac{1}{2}g_{\mu\nu}G = -\kappa T_{\mu\nu} \quad . \quad . \quad . \quad (1)$$

where $G_{\mu\nu}$ denotes the contracted Riemann tensor of curvature, G the scalar of curvature formed by repeated contraction, and $T_{\mu\nu}$ the energy-tensor of " matter." The assumption that the $T_{\mu\nu}$ do *not* depend on the derivatives of the $g_{\mu\nu}$ is in keeping with the historical development of these equations. For these quantities are, of course, the energy-components in the sense of the special theory of relativity, in which variable $g_{\mu\nu}$ do not occur. The second term on the left-hand side of the equation is so chosen that the divergence of the left-hand side of (1) vanishes identically, so that taking the divergence of (1), we obtain the equation

$$\frac{\partial \mathfrak{T}^{\sigma}_{\mu}}{\partial x_{\sigma}} + \tfrac{1}{2}g^{\sigma\tau}_{\mu}\mathfrak{T}_{\sigma\tau} = 0 \quad . \quad . \quad . \quad (2)$$

which in the limiting case of the special theory of relativity gives the complete equations of conservation

$$\frac{\partial T_{\mu\nu}}{\partial x_{\nu}} = 0.$$

Therein lies the physical foundation for the second term of the left-hand side of (1). It is by no means settled *a priori* that a limiting transition of this kind has any possible meaning. For if gravitational fields do play an essential part in the structure of the particles of matter, the transition to the limiting case of constant $g_{\mu\nu}$ would, for them, lose its justification, for indeed, with constant $g_{\mu\nu}$ there could not be any particles of matter. So if we wish to contemplate the possibility that gravitation may take part in the structure of the fields which constitute the corpuscles, we cannot regard equation (1) as confirmed.

Placing in (1) the Maxwell-Lorentz energy-components of the electromagnetic field $\phi_{\mu\nu}$,

$$T_{\mu\nu} = \tfrac{1}{4}g_{\mu\nu}\phi_{\sigma\tau}\phi^{\sigma\tau} - \phi_{\mu\sigma}\phi_{\nu\tau}g^{\sigma\tau}, \quad . \quad . \quad (3)$$

we obtain for (2), by taking the divergence, and after some reduction,*

$$\phi_{\mu\sigma}\mathfrak{J}^{\sigma} = 0 \quad . \quad . \quad . \quad . \quad (4)$$

where, for brevity, we have set

$$\frac{\partial}{\partial x_{\tau}}(\sqrt{-g}\ \phi_{\mu\nu}g^{\mu\sigma}g^{\nu\tau}) = \frac{\partial \mathfrak{f}^{\sigma\tau}}{\partial x_{\tau}} = \mathfrak{J}^{\sigma} \quad . \quad . \quad (5)$$

In the calculation we have employed the second of Maxwell's systems of equations

$$\frac{\partial \phi_{\mu\nu}}{\partial x_{\rho}} + \frac{\partial \phi_{\nu\rho}}{\partial x_{\mu}} + \frac{\partial \phi_{\rho\mu}}{\partial x_{\nu}} = 0 \quad . \quad . \quad . \quad (6)$$

We see from (4) that the current-density $\mathfrak{J}^{\sigma}$ must everywhere vanish. Therefore, by equation (1), we cannot arrive at a theory of the electron by restricting ourselves to the electromagnetic components of the Maxwell-Lorentz theory, as has long been known. Thus if we hold to (1) we are driven on to the path of Mie's theory.†

Not only the problem of matter, but the cosmological problem as well, leads to doubt as to equation (1). As I have shown in the previous paper, the general theory of relativity requires that the universe be spatially finite. But this view of the universe necessitated an extension of equations (1), with the introduction of a new universal constant λ, standing in a fixed relation to the total mass of the universe (or, respectively, to the equilibrium density of matter). This is gravely detrimental to the formal beauty of the theory.

§ 2. The Field Equations Freed of Scalars

The difficulties set forth above are removed by setting in place of field equations (1) the field equations

$$G_{\mu\nu} - \tfrac{1}{4}g_{\mu\nu}G = -\kappa T_{\mu\nu} \quad . \quad . \quad . \quad (1a)$$

where $T_{\mu\nu}$ denotes the energy-tensor of the electromagnetic field given by (3).

The formal justification for the factor $-\tfrac{1}{4}$ in the second

* Cf. e.g. A. Einstein, Sitzungsber. d. Preuss. Akad. d. Wiss., 1916, pp. 187, 188.

† Cf. D. Hilbert, Göttinger Nachr., 20 Nov., 1915.

term of this equation lies in its causing the scalar of the left-hand side,

$$g^{\mu\nu}(G_{\mu\nu} - \tfrac{1}{4}g_{\mu\nu}G),$$

to vanish identically, as the scalar $g^{\nu\mu}T_{\mu\nu}$ of the right-hand side does by reason of (3). If we had reasoned on the basis of equations (1) instead of (1a), we should, on the contrary, have obtained the condition $G = 0$, which would have to hold good everywhere for the $g_{\mu\nu}$, independently of the electric field. It is clear that the system of equations [(1a), (3)] is a consequence of the system [(1), (3)], but not conversely.

We might at first sight feel doubtful whether (1a) together with (6) sufficiently define the entire field. In a generally relativistic theory we need $n - 4$ differential equations, independent of one another, for the definition of n independent variables, since in the solution, on account of the liberty of choice of the co-ordinates, four quite arbitrary functions of all co-ordinates must naturally occur. Thus to define the sixteen independent quantities $g_{\mu\nu}$ and $\phi_{\mu\nu}$ we require twelve equations, all independent of one another. But as it happens, nine of the equations (1a), and three of the equations (6) are independent of one another.

Forming the divergence of (1a), and taking into account that the divergence of $G_{\mu\nu} - \frac{1}{2}g_{\mu\nu}G$ vanishes, we obtain

$$\phi_{\sigma\alpha}J^{\alpha} + \frac{1}{4\kappa}\frac{\partial G}{\partial x_{\sigma}} = 0 \quad . \quad . \quad . \quad (4a)$$

From this we recognize first of all that the scalar of curvature G in the four-dimensional domains in which the density of electricity vanishes, is constant. If we assume that all these parts of space are connected, and therefore that the density of electricity differs from zero only in separate "world-threads," then the scalar of curvature, everywhere outside these world-threads, possesses a constant value G_0. But equation (4a) also allows an important conclusion as to the behaviour of G within the domains having a density of electricity other than zero. If, as is customary, we regard electricity as a moving density of charge, by setting

$$J^{\sigma} = \frac{\mathfrak{J}^{\sigma}}{\sqrt{-g}} = \rho\frac{dx_{\sigma}}{ds}, \quad . \quad . \quad . \quad (7)$$

we obtain from (4a) by inner multiplication by J^σ, on account of the antisymmetry of $\phi_{\mu\nu}$, the relation

$$\frac{\partial G}{\partial x_\sigma}\frac{dx_\sigma}{ds} = 0 \quad . \quad . \quad . \quad . \quad (8)$$

Thus the scalar of curvature is constant on every world-line of the motion of electricity. Equation (4a) can be interpreted in a graphic manner by the statement: The scalar of curvature plays the part of a negative pressure which, outside of the electric corpuscles, has a constant value G_0. In the interior of every corpuscle there subsists a negative pressure (positive $G - G_0$) the fall of which maintains the electrodynamic force in equilibrium. The minimum of pressure, or, respectively, the maximum of the scalar of curvature, does not change with time in the interior of the corpuscle.

We now write the field equations (1a) in the form

$$(G_{\mu\nu} - \tfrac{1}{2}g_{\mu\nu}G) + \tfrac{1}{4}g_{\mu\nu}G_0 = -\kappa\left(T_{\mu\nu} + \frac{1}{4\kappa}g_{\mu\nu}(G - G_0)\right) \quad (9)$$

On the other hand, we transform the equations supplied with the cosmological term as already given

$$G_{\mu\nu} - \lambda g_{\mu\nu} = -\kappa(T_{\mu\nu} - \tfrac{1}{2}g_{\mu\nu}T).$$

Subtracting the scalar equation multiplied by $\frac{1}{2}$, we next obtain

$$(G_{\mu\nu} - \tfrac{1}{2}g_{\mu\nu}G) + g_{\mu\nu}\lambda = -\kappa T_{\mu\nu}.$$

Now in regions where only electrical and gravitational fields are present, the right-hand side of this equation vanishes. For such regions we obtain, by forming the scalar,

$$-G + 4\lambda = 0.$$

In such regions, therefore, the scalar of curvature is constant, so that λ may be replaced by $\frac{1}{4}G_0$. Thus we may write the earlier field equation (1) in the form

$$G_{\mu\nu} - \tfrac{1}{2}g_{\mu\nu}G + \tfrac{1}{4}g_{\mu\nu}G_0 = -\kappa T_{\mu\nu} \quad . \quad . \quad (10)$$

Comparing (9) with (10), we see that there is no difference between the new field equations and the earlier ones, except that instead of $T_{\mu\nu}$ as tensor of " gravitating mass " there now occurs $T_{\mu\nu} + \frac{1}{4\kappa} g_{\mu\nu}(G - G_0)$ which is independent of the scalar of curvature. But the new formulation has this great advantage, that the quantity λ appears in the fundamental equations as a constant of integration, and no longer as a universal constant peculiar to the fundamental law.

§ 3. On the Cosmological Question

The last result already permits the surmise that with our new formulation the universe may be regarded as spatially finite, without any necessity for an additional hypothesis. As in the preceding paper I shall again show that with a uniform distribution of matter, a spherical world is compatible with the equations.

In the first place we set

$$ds^2 = -\gamma_{ik}dx_i dx_k + dx_4^2 \; (i, k = 1, 2, 3) \qquad (11)$$

Then if P_{ik} and P are, respectively, the curvature tensor of the second rank and the curvature scalar in three-dimensional space, we have

$$\begin{aligned} G_{ik} &= P_{ik} \; (i, k = 1, 2, 3) \\ G_{i4} &= G_{4i} = G_{44} = 0 \\ G &= -P \\ -g &= \gamma. \end{aligned}$$

It therefore follows for our case that

$$\begin{aligned} G_{ik} - \tfrac{1}{2}g_{ik}G &= P_{ik} - \tfrac{1}{2}\gamma_{ik}P \; (i, k = 1, 2, 3) \\ G_{44} - \tfrac{1}{2}g_{44}G &= \tfrac{1}{2}P. \end{aligned}$$

We pursue our reflexions, from this point on, in two ways. Firstly, with the support of equation (1a). Here $T_{\mu\nu}$ denotes the energy-tensor of the electro-magnetic field, arising from the electrical particles constituting matter. For this field we have everywhere

$$\mathfrak{T}_1^1 + \mathfrak{T}_2^2 + \mathfrak{T}_3^3 + \mathfrak{T}_4^4 = 0.$$

The individual $\mathfrak{T}_\mu^\nu$ are quantities which vary rapidly with position; but for our purpose we no doubt may replace them by their mean values. We therefore have to choose

$$\left.\begin{aligned} \mathfrak{T}_1^1 = \mathfrak{T}_2^2 = \mathfrak{T}_3^3 &= -\tfrac{1}{3}\mathfrak{T}_4^4 = \text{const.} \\ \mathfrak{T}_\mu^\nu &= 0 \;\; (\text{for } \mu \neq \nu), \end{aligned}\right\} \quad . \quad . \quad (12)$$

and therefore

$$T_{ik} = \frac{1}{3}\frac{\mathfrak{T}_4^4}{\sqrt{\gamma}}\gamma_{ik},\ T_{44} = \frac{\mathfrak{T}_4^4}{\sqrt{\gamma}}.$$

In consideration of what has been shown hitherto, we obtain in place of (1a)

$$P_{ik} - \tfrac{1}{4}\gamma_{ik}P = -\tfrac{1}{3}\gamma_{ik}\frac{\kappa\mathfrak{T}_4^4}{\sqrt{\gamma}} \quad . \quad . \quad . \quad (13)$$

$$\tfrac{1}{4}P = -\frac{\kappa\mathfrak{T}_4^4}{\sqrt{\gamma}} \quad . \quad . \quad . \quad (14)$$

The scalar of equation (13) agrees with (14). It is on this account that our fundamental equations permit the idea of a spherical universe. For from (13) and (14) follows

$$P_{ik} + \frac{4}{3}\frac{\kappa\mathfrak{T}_4^4}{\sqrt{\gamma}}\gamma_{ik} = 0 \quad . \quad . \quad . \quad (15)$$

and it is known * that this system is satisfied by a (three-dimensional) spherical universe.

But we may also base our reflexions on the equations (9). On the right-hand side of (9) stand those terms which, from the phenomenological point of view, are to be replaced by the energy-tensor of matter; that is, they are to be replaced by

$$\begin{matrix} 0 & 0 & 0 & 0 \\ 0 & 0 & 0 & 0 \\ 0 & 0 & 0 & 0 \\ 0 & 0 & 0 & \rho \end{matrix}$$

where ρ denotes the mean density of matter assumed to be at rest. We thus obtain the equations

$$P_{ik} - \tfrac{1}{2}\gamma_{ik}P - \tfrac{1}{4}\gamma_{ik}G_0 = 0 \quad . \quad . \quad (16)$$

$$\tfrac{1}{2}P + \tfrac{1}{4}G_0 = -\kappa\rho \quad . \quad . \quad . \quad . \quad (17)$$

From the scalar of equation (16) and from (17) we obtain

$$G_0 = -\tfrac{2}{3}P = 2\kappa\rho, \quad . \quad . \quad . \quad (18)$$

and consequently from (16)

$$P_{ik} - \kappa\rho\gamma_{ik} = 0 \quad . \quad . \quad . \quad (19)$$

* Cf. H. Weyl, "Raum, Zeit, Materie," § 33.

which equation, with the exception of the expression for the co-efficient, agrees with (15). By comparison we obtain

$$\mathfrak{T}_4^4 = \tfrac{3}{4}\rho\sqrt{\gamma}. \qquad (20)$$

This equation signifies that of the energy constituting matter three-quarters is to be ascribed to the electromagnetic field, and one-quarter to the gravitational field.

§ 4. Concluding Remarks

The above reflexions show the possibility of a theoretical construction of matter out of gravitational field and electromagnetic field alone, without the introduction of hypothetical supplementary terms on the lines of Mie's theory. This possibility appears particularly promising in that it frees us from the necessity of introducing a special constant λ for the solution of the cosmological problem. On the other hand, there is a peculiar difficulty. For, if we specialize (1) for the spherically symmetrical static case we obtain one equation too few for defining the $g_{\mu\nu}$ and $\phi_{\mu\nu}$, with the result that *any spherically symmetrical distribution* of electricity appears capable of remaining in equilibrium. Thus the problem of the constitution of the elementary quanta cannot yet be solved on the immediate basis of the given field equations.

GRAVITATION AND ELECTRICITY

BY

H. WEYL

Translated from " Gravitation und Elektriticitat," Sitzungsberichte der Preussischen Akad. d. Wissenschaften, 1918.

GRAVITATION AND ELECTRICITY *

By H. WEYL

ACCORDING to Riemann,† geometry is based upon the following two facts :—

1. *Space is a Three-dimensional Continuum.*—The manifold of its points may therefore be consistently represented by the values of three co-ordinates x_1, x_2, x_3.

2. (*Pythagorean Theorem*).—The square of the distance ds between two infinitely proximate points

$$\mathrm{P} = (x_1, x_2, x_3) \text{ and } \mathrm{P}' = (x_1 + dx_1, x_2 + dx_2, x_3 + dx_3) \quad (1)$$

(any co-ordinates being employed) is a quadratic form of the relative co-ordinates dx_μ :—

$$ds^2 = \sum_{\mu\nu} g_{\mu\nu} dx_\mu dx_\nu, \quad (g_{\mu\nu} = g_{\nu\mu}) \qquad . \qquad . \quad (2)$$

The second of these facts may be briefly stated by saying that space is a *metrical* continuum. In complete accord with the spirit of the physics of immediate action we assume the Pythagorean theorem to be strictly valid only in the limit when the distances are infinitely small.

The special theory of relativity led to the discovery that *time* is associated as a fourth co-ordinate (x_4) on an equal footing with the three co-ordinates of space, and that the scene of material events, *the world*, is therefore *a four-dimensional, metrical continuum.* And so the quadratic form (2), which defines the metrical properties of the world, is not necessarily positive as in the case of the geometry of three-dimensional space, but has the index of inertia 3.‡ Riemann

* The footnotes in square brackets are later additions by the author.

† Math. Werke (2nd ed., Leipzig, 1892), No. XII, p. 282.

‡ That is to say that if the co-ordinates are chosen so that at one particular point of the continuum $ds^2 = \pm dx_1^2 \pm dx_2^2 \pm dx_3^2 \pm dx_4^3$, then in every case three of the signs will be + and one − (Trans.).

himself did not fail to point out that this quadratic form was to be regarded as a physical reality, since it reveals itself, e.g. in centrifugal forces, as the origin of real effects upon matter, and that matter therefore presumably reacts upon it. Until then all geometricians and philosophers had looked upon the metrical properties of space as pertaining to space itself, independently of the matter which it contained. It is upon this idea, which it was quite impossible for Riemann in his day to carry through, that Einstein in our own time, independently of Riemann, has raised the imposing edifice of his general theory of relativity. According to Einstein the phenomena of *gravitation* must also be placed to the account of geometry, and the laws by which matter affects measurements are no other than the laws of gravitation: the $g_{\mu\nu}$ in (2) form the components of the gravitational potential. While the gravitational potential thus consists of an invariant *quadratic* differential form, *electromagnetic phenomena* are governed by a four-potential of which the components ϕ_μ together compose an invariant *linear* differential form $\Sigma\phi_\mu dx_\mu$. But so far the two classes of phenomena, gravitation and electricity, stand side by side, the one separate from the other.

The later work of Levi-Civita,* Hessenberg,† and the author‡ shows quite plainly that the fundamental conception on which the development of Riemann's geometry must be based if it is to be in agreement with nature, is that of the infinitesimal parallel displacement of a vector. If P and P* are any two points connected by a curve, a given vector at P can be moved parallel to itself along this curve from P to P*. But, generally speaking, this conveyance of a vector from P to P* is not integrable, that is to say, the vector at P* at which we arrive depends upon the path along which the displacement travels. It is only in Euclidean "gravitationless" geometry that integrability obtains. The Riemannian geometry referred to above still contains a residual element of finite geometry—without any substantial reason, as far as I can see.

* "Nozione di parallelismo . . .", Rend. del Circ. Matem. di Palermo, Vol. 42 (1917).

† "Vektorielle Begründung der Differentialgeometrie," Math. Ann., Vol. 78 (1917).

‡ "Space, Time, and Matter" (1st ed., Berlin, 1918), § 14.

It seems to be due to the accidental origin of this geometry in the theory of surfaces. The quadratic form (2) enables us to compare, with respect to their length, not only two vectors at the same point, but also the vectors at any two points. *But a truly infinitesimal geometry must recognize only the principle of the transference of a length from one point to another point infinitely near to the first.* This forbids us to assume that the problem of the transference of length from one point to another at a finite distance is integrable, more particularly as the problem of the transference of direction has proved to be non-integrable. Such an assumption being recognized as false, a geometry comes into being, which, when applied to the world, explains in a surprising manner *not only the phenomena of gravitation, but also those of the electromagnetic field.* According to the theory which now takes shape, both classes of phenomena spring from the same source, and in fact *we cannot in general make any arbitrary separation of electricity from gravitation.* In this theory *all physical quantities have a meaning in world geometry.* In particular the quantities denoting physical effects appear at once as pure numbers. The theory leads to a world-law which in its essentials is defined without ambiguity. It even permits us in a certain sense to comprehend why the world has four dimensions. I shall now first of all give a sketch of the structure of the amended geometry of Riemann without any thought of its physical interpretation. Its application to physics will then follow of its own accord.

In a given system of co-ordinates the relative co-ordinates dx_μ of a point P infinitely near to P—see (1)—are the components of the infinitesimal displacement PP′. The transition from one system of co-ordinates to another is expressed by definite formulæ of transformation,

$$x_\mu = x_\mu(x_1^*, x_2^* \ldots x_n^*) \quad \mu = 1, 2, \ldots n,$$

which determine the connexion between the co-ordinates of the same point in the two systems. Then between the components dx_μ and the components dx_μ^* of the same infinitesimal displacement of the point P we have the linear formulæ of transformation

$$dx_\mu = \sum_\nu a_{\mu\nu} dx_\nu^* \quad . \quad . \quad . \quad . \quad (3)$$

in which $a_{\mu\nu}$ are the values of the derivatives $\frac{\partial x_\mu}{\partial x^*_\nu}$ at the point P. A contravariant vector **x** at the point P referred to either system of co-ordinates has n known numbers ξ^μ for its components, which in the transition to another system are transformed in exactly the same way (3) as the components of an infinitesimal displacement. I denote the totality of vectors at the point P as the vector-space at P. It is, firstly, linear or affine, i.e. by multiplication of a vector at P by a number, and by addition of two such vectors, there always arises a vector at P; and, secondly, it is metrical, i.e. by the symmetrical bilinear form belonging to (2) a scalar product

$$\mathbf{x}\,.\,\mathbf{y} = \mathbf{y}\,.\,\mathbf{x} = \sum_{\mu\nu} g_{\mu\nu}\xi^\mu\eta^\nu$$

is invariantly assigned to each pair of vectors **x** and **y** with components ξ^μ, η^μ. We take it, however, that this form is determined only as far as to a positive factor of proportionality, which remains arbitrary. If the manifold of points of space is represented by co-ordinates x_μ, the $g_{\mu\nu}$ are determined by the metrical properties at the point P only to the extent of their proportionality. In the physical sense, too, it is only the ratios of the $g_{\mu\nu}$ that has an immediate tangible meaning. For the equation

$$\sum_{\mu\nu} g_{\mu\nu}dx_\mu dx_\nu = 0$$

is satisfied, when P is a given origin, by those infinitely proximate world-points which are reached by a light signal emitted at P. For the purpose of analytical presentation we have firstly to choose a definite system of co-ordinates, and secondly at each point P to determine the arbitrary factor of proportionality with which the $g_{\mu\nu}$ are endowed. Accordingly the formulæ which emerge must possess a double property of invariance: they must be invariant with respect to any continuous transformations of co-ordinates, and they must remain unaltered if $\lambda g_{\mu\nu}$, where λ is an arbitrary continuous function of position, is substituted for the $g_{\mu\nu}$. The supervention of this second property of invariance is characteristic of our theory.

If P, P* are any two points, and if to each vector **x** at P a vector **x*** at P* is assigned in such a way that in general $a\mathbf{x}$

becomes $a\mathbf{x}^*$, and $\mathbf{x}+\mathbf{y}$ becomes $\mathbf{x}^*+\mathbf{y}^*$ (a being any assigned number), and the vector zero at P is the only one to which the vector zero at P* corresponds, we then have made an affine or linear replica of the vector-space at P on the vector-space at P*. This replica has a particularly close resemblance when the scalar product of the vectors $\mathbf{x}^*$, $\mathbf{y}^*$ at P* is proportional to that of $\mathbf{x}$ and $\mathbf{y}$ at P for all pairs of vectors $\mathbf{x}$, $\mathbf{y}$. (In our view it is only this idea of a similar replica that has an objective sense; the previous theory permitted the more definite conception of a congruent replica.) The meaning of the parallel displacement of a vector at the point P to a neighbouring point P′ is settled by the two axiomatic postulates.

1. By the parallel displacement of the vectors at the point P to the neighbouring point P′ a similar image of the vector-space at P is made upon the vector-space at P′.

2. If P_1, P_2 are two points in the neighbourhood of P, and the infinitesimal vector PP_2 at P is transformed into P_1P_{12} by a parallel displacement to the point P_1, while PP_1 at P is transformed into P_2P_{21} by parallel displacement to P_2, then P_{12}, P_{21} coincide, i.e. infinitesimal parallel displacements are commutative.

That part of postulate 1 which says that the parallel displacement is an affine transposition of the vector-space from P to P′, is expressed analytically as follows: the vector ξ^μ at $P=(x_1, x_2, \ldots x_n)$ is by displacement transformed into a vector $\xi^\mu + d\xi^\mu$ at $P' = (x_1+dx_1, x_2+dx_2, \ldots x_n+dx_n)$ the components of which are in a linear relation to ξ^μ,—

$$d\xi^\mu = -\sum_\nu d\gamma^\mu_\nu \xi^\nu \quad . \quad . \quad . \quad . \quad (4)$$

The second postulate teaches that the $d\gamma^\mu_\nu$ are linear differential forms

$$d\gamma^\mu_\nu = \sum_\rho \Gamma^\mu_{\nu\rho} dx_\rho,$$

the coefficients of which possess the symmetrical property

$$\Gamma^\mu_{\nu\rho} = \Gamma^\mu_{\rho\nu} \quad . \quad . \quad . \quad . \quad (5)$$

If two vectors ξ^μ, η^μ at P are transformed by parallel displacement at P′ into $\xi^\mu + d\xi^\mu$, $\eta^\mu + d\eta^\mu$, then the postulate of similarity stated under 1 above, which goes beyond affinity, tells us that

$$\sum_{\mu\nu}(g_{\mu\nu} + dg_{\mu\nu})(\xi^\mu + d\xi^\mu)(\eta^\nu + d\eta^\nu)$$

must be proportional to

$$\sum_{\mu\nu} g_{\mu\nu}\xi^\mu\eta^\nu.$$

If we call the factor of proportionality, which differs infinitesimally from 1, $1 + d\phi$, and define the reduction of an index in the usual way by the formula

$$a_\mu = \sum_\nu g_{\mu\nu}a^\nu,$$

we obtain

$$dg_{\mu\nu} - (d\gamma_{\nu\mu} + d\gamma_{\mu\nu}) = g_{\mu\nu}d\phi \quad . \quad . \quad (6)$$

From this it follows that $d\phi$ is a linear differential form

$$d\phi = \sum_\mu \phi_\mu dx_\mu \quad . \quad . \quad . \quad . \quad (7)$$

If this is known, the equation (6) or

$$\Gamma_{\mu,\nu\rho} + \Gamma_{\nu,\mu\rho} = \frac{\partial g_{\mu\nu}}{\partial x_\rho} - g_{\mu\nu}\phi_\rho,$$

together with the condition for symmetry (5), gives unequivocally the quantities Γ. *The internal metrical connexion of space thus depends on a linear form* (7) *besides the quadratic form* (2)—which is determined except as to an arbitrary factor of proportionality.* If we substitute $\lambda g_{\mu\nu}$ for $g_{\mu\nu}$ with-

* [I have now modified this structure in the following points (cf. the final presentation in ed. 4 of "Raum, Zeit, Materie," 1921, §§ 13, 18). (*a*) In place of postulates 1 and 2, which the parallel displacement has to fulfil, there is now one postulate: Let there be a system of co-ordinates at the point P, by the employment of which the components of every vector at P are not altered by parallel displacement to any point in infinite proximity to P. This postulate characterizes the essence of the parallel displacement as that of a transposition, concerning which it may be correctly asserted that it leaves the vectors "unaltered." (*b*) To the metrics at the single point P, according to which there is attached to every vector $\mathbf{x} = \xi^\mu$ at P a tract of such a kind that two vectors define the same tract when, and only when, they possess the same measure-number $l = \sum g_{\mu\nu}\xi^\mu\xi^\nu$, there must now be added the metrical connexion of P with the points in its neighbourhood: by congruent transposition to the infinitely near point P′ a tract at P passes over into a definite tract at P′. If we make a requirement of this concept of congruent transposition of tracts analogous to that which has just been postulated, under (*a*), of the concept of parallel displacement of vectors, we see that this process (in which the measure-number l of the tract is increased by dl) is expressed in the equations

$$dl = l d\phi\,;\ d\phi = \sum \phi_\mu dx_\mu.$$

out changing the system of co-ordinates, the quantities $d\gamma^{\mu}_{\nu}$ do not change, $d\gamma_{\mu\nu}$ assumes the factor λ, and $dg_{\mu\nu}$ becomes $\lambda dg_{\mu\nu} + g_{\mu\nu}d\lambda$. Equation (6) then shows that $d\phi$ becomes

$$d\phi + \frac{d\lambda}{\lambda} = d\phi + d\,(\log \lambda).$$

What remains undetermined, therefore, in the linear form $\Sigma\phi_{\mu}dx_{\mu}$ is not a factor of proportionality which would have to be settled by an arbitrary choice of a unit of measurement, but, rather, the arbitrary element inherent in it consists in an additive total differential. For the analytical representation of geometry the forms

$$g_{\mu\nu}dx_{\mu}dx_{\nu}, \quad \phi_{\mu}dx_{\mu} \; . \qquad . \qquad . \qquad . \qquad (8)$$

are on an equal footing with

$$\lambda \, . \, g_{\mu\nu}dx_{\mu}dx_{\nu} \text{ and } \phi_{\mu}dx_{\mu} + d\,(\log \lambda) \quad . \qquad . \qquad (9)$$

where λ is any positive function of position. Hence there is invariant significance in the anti-symmetrical tensor with the components

$$F_{\mu\nu} = \frac{\partial\phi_{\mu}}{\partial x_{\nu}} - \frac{\partial\phi_{\nu}}{\partial x_{\mu}} \qquad . \qquad . \qquad . \qquad . \qquad (10)$$

i.e. the form

$$F_{\mu\nu} = dx_{\mu}\delta x_{\nu} = \tfrac{1}{2}F_{\mu\nu}\Delta x_{\mu\nu}$$

which depends bilinearly on two arbitrary displacements dx and δx at the point P—or, rather, depends linearly on the surface element with the components $\Delta x_{\mu\nu} = dx_{\mu}\delta x_{\nu} - dx_{\nu}\delta x_{\mu}$ which is defined by these two displacements. The special case of the theory as hitherto developed, in which the arbitrarily chosen unit of length at the origin allows itself to be transferred by parallel displacement to all points of space in a manner which is independent of the path traversed —this special case occurs when the $g_{\mu\nu}$ can be absolutely determined in such a way that the ϕ_{μ} vanish. The $\Gamma^{\mu}_{\nu\rho}$ are

In these circumstances the metrics and the metrical connexion determine the "affine" connexion (parallel displacement) without ambiguity—and indeed, according to my present view of the problem of space this is the most fundamental fact of geometry—whereas according to the presentation given in the text it is the linear form $d\phi$ that remains arbitrary in the given metrics at the parallel displacement.]

then nothing else than the Christoffel three-indices symbols. The necessary and sufficient invariant condition for the occurrence of this case consists in the identical vanishing of the tensor $F_{\mu\nu}$.

This naturally suggests interpreting ϕ_μ in world-geometry as the four-potential, and the tensor F consequently as electromagnetic field. For the absence of an electromagnetic field is the necessary condition for the validity of Einstein's theory, which, up to the present, accounts for the phenomena of gravitation only. If this view is accepted, it will be seen that the electric quantities are of such a nature that their characterization by numbers in a definite system of co-ordinates does not depend on the arbitrary choice of a unit of measurement. In fact, in the question of the unit of measurement and of dimension there must be a new orientation of the theory. Hitherto a quantity has been spoken of as, e.g., a tensor of the second rank, if a single value of the quantity determines a matrix of numbers $a_{\mu\nu}$ in each system of coordinates after an arbitrary unit of measurement has been selected, these numbers forming the coefficients of an invariant bilinear form of two arbitrary, infinitesimal displacements

$$a_{\mu\nu} dx_\mu \delta x_\nu \quad . \quad . \quad . \quad . \quad (11)$$

But here we speak of a tensor, if, with a system of co-ordinates taken as a base, and after definite selection of the factor of proportionality contained in the $g_{\mu\nu}$, the components $a_{\mu\nu}$ are determined without ambiguity and in such a way that on transforming the co-ordinates the form (11) remains invariant, but on replacing $g_{\mu\nu}$ by $\lambda g_{\mu\nu}$ the $a_{\mu\nu}$ become $\lambda^e a_{\mu\nu}$. We then say that the tensor has the weight e, or, ascribing to the linear element ds the dimension "length $= l$," that it is of dimension l^{2e}. Only those tensors of weight 0 are absolutely invariant. The field tensor with the components $F_{\mu\nu}$ is of this kind. By (10) it satisfies the first system of the Maxwell equations

$$\frac{\partial F_{\nu\rho}}{\partial x_\mu} + \frac{\partial F_{\rho\mu}}{\partial x_\nu} + \frac{\partial F_{\mu\nu}}{\partial x_\rho} = 0.$$

When once the idea of parallel displacement is clear, geometry and the tensor calculus can be established without difficulty.

(*a*) *Geodetic Lines.*—Given a point P and at that point a vector, the geodetic line from P in the direction of this vector is given by continuously moving the vector parallel to itself in its own direction. Employing a suitable parameter τ the differential equation of the geodetic line is

$$\frac{d^2x_\mu}{d\tau^2} + \Gamma^\mu_{\nu\rho}\frac{dx_\nu}{d\tau}\frac{dx_\rho}{d\tau} = 0.$$

(Of course it cannot be characterized as the line of smallest length, because the notion of curve-length has no meaning.)

(*b*) *Tensor Calculus.*—To deduce, for example, a tensor field of rank 2 by differentiation from a covariant tensor field of rank 1 and weight 0 with components f_μ, we call in the help of an arbitrary vector ξ^μ at the point P, form the invariant $f_\mu\xi^\mu$ and its infinitely small alteration on transition from the point P with the co-ordinates x_μ to the neighbouring point P′ with the co-ordinates $x_\mu + dx_\mu$ by shifting the vector along a parallel to itself during this transition. For this alteration we have

$$\frac{\partial f_\mu}{\partial x_\nu}\xi^\mu dx_\nu + f_\rho d\xi^\rho = \left(\frac{\partial f_\mu}{\partial x_\nu} - \Gamma^\rho_{\mu\nu}f_\rho\right)\xi^\mu dx_\nu.$$

The quantities in brackets on the right are therefore the components of a tensor field of rank 2 and weight 0, which is formed from the field f in a perfectly invariant manner.

(*c*) *Curvature.*—To construct the analogue to Riemann's tensor of curvature, let us begin with the figure employed above, of an infinitely small parallelogram, consisting of the points P, P_1, P_2, and $P_{12} = P_{21}$.* If we displace a vector $\mathbf{x} = \xi^\mu$ at P parallel to itself, to P_1 and from there to P_{12}, and a second time first to P_2 and thence to P_{21}, then, since P_{12} and P_{21} coincide, there is a meaning in forming the difference $\Delta\mathbf{x}$ of the two vectors obtained at this point. For their components we have

$$\Delta\xi^\mu = \Delta R^\mu_\nu \xi^\nu \quad . \quad . \quad . \quad . \quad (12)$$

where the ΔR^μ_ν are independent of the displaced vector $\mathbf{x}$, but

* [Here it is not essential that opposite sides of the infinitely small "parallelogram" are produced by parallel displacement one from the other; we are concerned only with the coincidence of the points P_{12} and P_{21}.]

on the other hand depend linearly on the surface-element defined by the two displacements $PP_1 = dx_\mu$, $PP_2 = \delta x_\mu$. Thus

$$\Delta R^\mu_\nu = R^\mu_{\nu\rho\sigma} dx_\rho \delta x_\sigma = \tfrac{1}{2} R^\mu_{\nu\rho\sigma} \Delta x_{\rho\sigma}.$$

The components of curvature $R^\mu_{\nu\rho\sigma}$, depending solely on the place P, possess the two properties of symmetry that (1) they change sign on the interchange of the last two indices ρ and σ, and (2), if we perform the three cylic interchanges $\nu\rho\sigma$, and add up the appropriate components, the result is 0. Reducing the index μ, we obtain at $R_{\mu\nu\rho\sigma}$ the components of a covariant tensor of rank 4 and weight 1. Even without calculation we see that R divides in a natural, invariant manner into two parts,

$$R^\mu_{\nu\rho\sigma} = P^\mu_{\nu\rho\sigma} - \tfrac{1}{2}\delta^\mu_\nu F_{\rho\sigma} \quad (\delta^\mu_\nu = 1 \text{ if } \mu = \nu\,; \; = 0 \text{ if } \mu \neq \nu), \quad (13)$$

of which the first, $P^\mu_{\nu\rho\sigma}$, is anti-symmetrical, not only in the indices $\rho\sigma$, but also in μ and ν. Whereas the equations $F_{\mu\nu} = 0$ characterize our space as one without an electromagnetic field, i.e. as one in which the problem of the conveyance of length is integrable, the equations $P^\mu_{\nu\rho\sigma} = 0$ are, as (13) shows, the invariant conditions for the absence of a gravitational field, i.e. for the problem of the conveyance of direction to be integrable. The Euclidean space alone is one which at the same time is free of electricity and of gravitation.

The simplest invariant of a linear copy like (12), which to each vector **x** assigns a vector Δ**x**, is its "spur"

$$\frac{1}{n}\Delta R^\mu_\mu.$$

For this, by (13), we obtain in the present case the form

$$-\tfrac{1}{2} F_{\rho\sigma} dx_\rho \delta x_\sigma$$

which we have already encountered above. The simplest invariant of a tensor like $-\tfrac{1}{2}F_{\rho\sigma}$ is the "square of its magnitude"

$$L = \tfrac{1}{4} F_{\rho\sigma} F^{\rho\sigma} \quad . \quad . \quad . \quad . \quad (14)$$

L is evidently an invariant of weight -2, because the tensor F has weight 0. If g is the negative determinant of the $g_{\mu\nu}$, and

$$d\omega = \sqrt{g}\, dx_0 dx_1 dx_2 dx_3 = \sqrt{g}\, dx$$

the volume of an infinitely small element of volume, it is known that the Maxwell theory is governed by the quantity of electrical action, which is equal to the integral $\int L d\omega$ of this simplest invariant, extended over any chosen territory, and indeed is governed in the sense that, with any variations of the $g_{\mu\nu}$ and ϕ_μ, which vanish at the limits of world-territory, we have

$$\delta \int L d\omega = \int (S^\mu d\phi_\mu + T^{\mu\nu} \delta g_{\mu\nu}) d\omega,$$

where

$$S^\mu = \frac{1}{\sqrt{g}} \frac{\partial(\sqrt{g} F^{\mu\nu})}{\partial x_\nu}$$

are the left-hand sides of the generalized Maxwellian equations (the right-hand sides of which are the components of the four-current), and the $T^{\mu\nu}$ form the energy-momentum tensor of the electromagnetic field. As L is an invariant of weight -2, whereas the volume-element in n-dimensional geometry is an invariant of weight $\frac{1}{2}n$, the integral has significance only when the number of dimensions $n = 4$. Thus on our interpretation the possibility of the Maxwell theory is restricted to the case of four dimensions. In the four-dimensional world, however, the quantity of electromagnetic action becomes a pure number. Nevertheless, the magnitude of the quantity 1 cannot be ascertained in the traditional units of the c.g.s. system until a physical problem, to be tested by observation (as for example the electron), has been calculated on the basis of our theory.

Passing now from geometry to physics, we have to assume, following the precedent of Mie's theory,* that all the laws of nature rest upon a definite integral invariant, the action-quantity

$$\int W d\omega = \int \mathfrak{W} dx, \quad \mathfrak{W} = W\sqrt{g},$$

in such a way that the real world is distinguished from all other possible four-dimensional metrical spaces by the characteristic that for it the action-quantity contained in any part of its domain assumes a stationary value in relation to such variations of the potentials $g_{\mu\nu}$, ϕ_μ as vanish at the limits of

* Ann. d. Physik, 37, 39, 40, 1912-13.

the territory in question. W, the world-density of the action, must be an invariant of weight -2. The action-quantity is in any case a pure number; thus our theory at once accounts for that atomistic structure of the world to which current views attach the most fundamental importance—the action-quantum. The simplest and most natural conjecture which we can make for W, is

$$W = R^{\mu}_{\nu\rho\sigma} R^{\nu\rho\sigma}_{\mu} = |R|^2.$$

For this we also have, by (13),

$$W = |P|^2 + 4L.$$

(There could be no doubt about anything here except perhaps the factor 4, with which the electric term L is added to the first.) But even without particularizing the action-quantity we can draw some general conclusions from the principle of action. For we shall show that as, according to investigations by Hilbert, Lorentz, Einstein, Klein, and the author,* the four laws of the conservation of matter (the energy-momentum tensor) are connected with the invariance of the action quantity (containing four arbitrary functions) with respect to transformations of co-ordinates, so in the same way the law of the conservation of electricity is connected with the "measure-invariance" [transition from (8) to (9)] which here makes its appearance for the first time, introducing a fifth arbitrary function. The manner in which the latter associates itself with the principles of energy and momentum seems to me one of the strongest general arguments in favour of the theory here set out—so far as there can be any question at all of confirmation in purely speculative matters.

For any variation which vanishes at the limits of the world-territory under consideration we have

$$\delta\int \mathfrak{W} dx = \int (\mathfrak{W}^{\mu\nu}\delta g_{\mu\nu} + \mathfrak{w}^{\mu}\delta\phi_{\mu})dx \quad (\mathfrak{W}^{\mu\nu} = \mathfrak{W}^{\mu\nu}) \quad (15)$$

* Hilbert, "Die Grundlagen der Physik," Göttinger Nachrichten, 20 Nov., 1915; H. A. Lorentz in four papers in the Versl. K. Ak. van Wetensch., Amsterdam, 1915-16; A. Einstein, Berl. Ber., 1916, pp. 1111-6; F. Klein, Gött. Nachr., 25 Jan., 1918; H. Weyl, Ann. d. Physik, 54, 1917, pp. 121-5.

The laws of nature then take the form

$$\mathfrak{W}^{\mu\nu} = 0, \mathfrak{w}^{\mu} = 0 \quad . \quad . \quad . \quad (16)$$

The former may be regarded as the laws of the gravitational field, the latter as those of the electromagnetic field. The quantities W^{μ}_{ν}, w^{μ} defined by

$$\mathfrak{W}^{\mu}_{\nu} = \sqrt{g}W^{\mu}_{\nu}, \quad \mathfrak{w}^{\mu} = \sqrt{g}w^{\mu}$$

are the mixed or, respectively, the contravariant components of a tensor of rank 2 or 1 respectively, and of weight − 2. In the system of equations (16) there are five which are redundant, in accordance with the properties of invariance. This is expressed in the following five invariant identities, which subsist between their left-hand sides :—

$$\frac{\partial \mathfrak{w}^{\mu}}{\partial x_{\mu}} \equiv \mathfrak{W}^{\mu}_{\mu} \quad . \quad . \quad . \quad . \quad . \quad (17)$$

$$\frac{\partial \mathfrak{W}^{\mu}_{\nu}}{\partial x_{\mu}} - \Gamma^{\alpha}_{\nu\beta}\mathfrak{W}^{\beta}_{\alpha} \equiv \tfrac{1}{2}F_{\mu\nu}\mathfrak{w}^{\mu} \quad . \quad . \quad . \quad (18)$$

The first results from the measure-invariance. For if in the transition from (8) to (9) we assume for log λ an infinitely small function of position $\delta\rho$, we obtain the variation

$$\delta g_{\mu\nu} = g_{\mu\nu}\delta\rho, \quad \delta\phi_{\mu} = \frac{\partial(\delta\rho)}{\partial x_{\mu}}.$$

For this variation (15) must vanish. In the second place if we utilize the invariance of the action-quantity with respect to transformations of co-ordinates by means of an infinitely small deformation of the world - continuum,* we obtain the identities

$$\frac{\partial \mathfrak{W}^{\mu}_{\nu}}{\partial x_{\mu}} - \frac{1}{2}\frac{\partial g_{\alpha\beta}}{\partial x_{\nu}}\mathfrak{W}^{\alpha\beta} + \frac{1}{2}\left(\frac{\partial \mathfrak{w}^{\mu}}{\partial x_{\mu}}\phi_{\nu} - \Gamma_{\alpha\nu}\mathfrak{w}^{\alpha}\right) \equiv 0$$

which change into (18) when, by (17) $\partial\mathfrak{w}^{\mu}/\partial x_{\mu}$ is replaced by $g_{\alpha\beta}\mathfrak{W}^{\alpha\beta}$

From the gravitational laws alone therefore we already obtain

$$\frac{\partial \mathfrak{w}^{\mu}}{\partial x_{\mu}} = 0, \quad . \quad . \quad . \quad . \quad (19)$$

* Weyl, Ann. d. Physik, 54, 1917, pp. 121-5; F. Klein, Gött. Nachr., 25 Jan., 1918.

and from the laws of the electromagnetic field alone

$$\frac{\partial}{\partial x_\mu}\mathfrak{W}^\mu_\nu - \Gamma^\alpha_{\nu\beta}\mathfrak{W}^\beta_\alpha = 0 \quad . \quad . \quad . \quad (20)$$

In Maxwell's theory $\mathfrak{w}^\mu$ has the form

$$\mathfrak{w}^\mu \equiv \frac{\partial(\sqrt{g}F^{\mu\nu})}{\partial x_\nu} - \mathfrak{s}^\mu, \quad \mathfrak{s}^\mu = \sqrt{g}s^\mu$$

where s^μ denotes the four-current. Since the first part here satisfies the equation (19) identically, this equation gives us the law of conservation of electricity

$$\frac{1}{\sqrt{g}}\frac{\partial(\sqrt{g}s^\mu)}{\partial x_\mu} = 0.$$

Similarly in Einstein's theory of gravitation $\mathfrak{W}^\mu_\nu$ consists of two terms, the first of which satisfies equation (20) identically, and the second is equal to the mixed components of the energy-momentum tensor T^μ_ν multiplied by $\sqrt{g}$. Thus equations (20) lead to the four laws of the conservation of matter. Quite analogous circumstances hold good in our theory if we choose the form (14) for the action-quantity. The five principles of conservation are "eliminants" of the field laws, i.e. they follow from them in a twofold manner, and thus demonstrate that among them there are five which are redundant.

With the form (14) for the action-quantity the Maxwell equations run, for example :—

$$\frac{1}{\sqrt{g}}\frac{\partial(\sqrt{g}F^{\mu\nu})}{\partial x_\nu} = s^\mu, \quad . \quad . \quad . \quad (21)$$

and the current is

$$s_\mu = \tfrac{1}{4}\left(R\phi_\mu + \frac{\partial R}{\partial x_\mu}\right),$$

where R denotes that invariant of weight -1 which arises from $R^\mu_{\nu\rho\sigma}$ if we first contract with respect to μ, ρ and then with respect to ν and σ. If R^* denotes Riemann's invariant of curvature constructed solely from the $g_{\mu\nu}$, calculation gives

$$R = R^* - \frac{3}{\sqrt{g}} \frac{\partial(\sqrt{g}\phi^\mu)}{\partial x_\mu} + \frac{3}{2}\phi_\mu\phi^\mu.$$

In the static case, where the space components of the electromagnetic potential disappear, and all quantities are independent of the time x_0, by (21) we must have

$$R = R^* + \frac{3}{2}\phi_0\phi^0 = \text{const.}$$

But in a world-territory in which $R \neq 0$ we may make $R = \text{const.} = \pm 1$ everywhere, by appropriate determination of the unit of length. Only we have to expect, under conditions which are variable with time, surfaces $R = 0$, which evidently will play some singular part. R cannot be used as density of action (represented by R^* in Einstein's theory of gravitation) because it has not the weight -2. The consequence is that though our theory leads to Maxwell's electromagnetic equations, it does not lead to Einstein's gravitation equations. In their place appear differential equations of order 4. But indeed it is very improbable that Einstein's equations of gravitation are strictly correct, because, above all things, the gravitation constant occurring in them is not at all in the picture with the other constants of nature, the gravitation radius of the charge and mass of an electron, for example, being of an entirely different order of magnitude (10^{20} or 10^{40} times as small) from that of the radius of the electron itself.*

It was my intention here merely to develop briefly the general principles of the theory.† The problem naturally

* Cf. Weyl, Ann. d. Physik, 54, 1917, p. 133.

† [The problem of defining all W invariants allowable as action-quantities, under the requirement that they should contain the derivatives of the $g_{\mu\nu}$ only to the second order at most, and those of the ϕ_μ only to the first order, was solved by R. Weitzenböck (Sitzungsber. d. Akad. d. Wissensch. in Wien, 129, 1920; 130, 1921). If we omit the invariants W for which the variation $\delta\int W d\omega$ vanishes identically, there remain according to a later calculation by R. Bach (Math. Zeitschrift, 9, 1921, pp. 125 and 189) only three combinations. The real W seems to be a linear combination of Maxwell's L and the square of R. This conjecture has been tested more carefully by W. Pauli (Physik. Zeitschrift., 20, 1919, pp. 457-67) and myself; in particular we succeeded in advancing so far on this basis as to deduce the equations of

presents itself of deducing the physical consequences of the theory on the basis of the special form for the action-quantity given in (14), and of comparing these with experience, examining particularly whether the existence of the electron and the peculiarities of the hitherto unexplained processes in the atom can be deduced from the theory.* The task is extraordinarily complicated from the mathematical point of view, because it is impossible to obtain approximate solutions if we restrict ourselves to the linear terms; for since it is certainly not permissible to neglect terms of higher order in the interior of the electron, the linear equations obtained by neglecting these may have, in general, only the solution 0. I propose to return to all these matters in greater detail in another place.

motion of a material particle. The invariant (14) selected above, at hazard in the first place, seems on the contrary to play no part in nature. Cf. Raum, "Zeit, Materie," ed. 4, §§ 35, 36, or Weyl, Physik. Zeitschr., 22, 1921, pp. 473-80.]

* [Meanwhile I have quite abandoned these hopes, raised by Mie's theory; I do not believe that the problem of matter is to be solved by a mere field theory. Cf. on this subject my article "Feld und Materie," Ann. d. Physik, 65, 1921, pp. 541-63.]

A CATALOG OF SELECTED DOVER BOOKS IN ALL FIELDS OF INTEREST

A CATALOG OF SELECTED DOVER BOOKS IN ALL FIELDS OF INTEREST

THE MURDER BOOK OF J.G. REEDER, Edgar Wallace. Eight suspenseful stories by bestselling mystery writer of 20s and 30s. Features the donnish Mr. J.G. Reeder of Public Prosecutor's Office. 128pp. 5⅜ × 8½. (Available in U.S. only) 24374-5 Pa. $3.50

ANNE ORR'S CHARTED DESIGNS, Anne Orr. Best designs by premier needlework designer, all on charts: flowers, borders, birds, children, alphabets, etc. Over 100 charts, 10 in color. Total of 40pp. 8¼ × 11. 23704-4 Pa. $2.25

BASIC CONSTRUCTION TECHNIQUES FOR HOUSES AND SMALL BUILDINGS SIMPLY EXPLAINED, U.S. Bureau of Naval Personnel. Grading, masonry, woodworking, floor and wall framing, roof framing, plastering, tile setting, much more. Over 675 illustrations. 568pp. 6½ × 9¼. 20242-9 Pa. $8.95

MATISSE LINE DRAWINGS AND PRINTS, Henri Matisse. Representative collection of female nudes, faces, still lifes, experimental works, etc., from 1898 to 1948. 50 illustrations. 48pp. 8⅜ × 11¼. 23877-6 Pa. $2.50

HOW TO PLAY THE CHESS OPENINGS, Eugene Znosko-Borovsky. Clear, profound examinations of just what each opening is intended to do and how opponent can counter. Many sample games. 147pp. 5⅜ × 8½. 22795-2 Pa. $2.95

DUPLICATE BRIDGE, Alfred Sheinwold. Clear, thorough, easily followed account: rules, etiquette, scoring, strategy, bidding; Goren's point-count system, Blackwood and Gerber conventions, etc. 158pp. 5⅜ × 8½. 22741-3 Pa. $3.00

SARGENT PORTRAIT DRAWINGS, J.S. Sargent. Collection of 42 portraits reveals technical skill and intuitive eye of noted American portrait painter, John Singer Sargent. 48pp. 8¼ × 11⅛. 24524-1 Pa. $2.95

ENTERTAINING SCIENCE EXPERIMENTS WITH EVERYDAY OBJECTS, Martin Gardner. Over 100 experiments for youngsters. Will amuse, astonish, teach, and entertain. Over 100 illustrations. 127pp. 5⅜ × 8½. 24201-3 Pa. $2.50

TEDDY BEAR PAPER DOLLS IN FULL COLOR: A Family of Four Bears and Their Costumes, Crystal Collins. A family of four Teddy Bear paper dolls and nearly 60 cut-out costumes. Full color, printed one side only. 32pp. 9¼ × 12¼. 24550-0 Pa. $3.50

NEW CALLIGRAPHIC ORNAMENTS AND FLOURISHES, Arthur Baker. Unusual, multi-useable material: arrows, pointing hands, brackets and frames, ovals, swirls, birds, etc. Nearly 700 illustrations. 80pp. 8⅜ × 11¼. 24095-9 Pa. $3.50

DINOSAUR DIORAMAS TO CUT & ASSEMBLE, M. Kalmenoff. Two complete three-dimensional scenes in full color, with 31 cut-out animals and plants. Excellent educational toy for youngsters. Instructions; 2 assembly diagrams. 32pp. 9¼ × 12¼. 24541-1 Pa. $3.95

SILHOUETTES: A PICTORIAL ARCHIVE OF VARIED ILLUSTRATIONS, edited by Carol Belanger Grafton. Over 600 silhouettes from the 18th to 20th centuries. Profiles and full figures of men, women, children, birds, animals, groups and scenes, nature, ships, an alphabet. 144pp. 8⅜ × 11¼. 23781-8 Pa. $4.50

25 KITES THAT FLY, Leslie Hunt. Full, easy-to-follow instructions for kites made from inexpensive materials. Many novelties. 70 illustrations. 110pp. 5⅜ × 8½. 22550-X Pa. $1.95

PIANO TUNING, J. Cree Fischer. Clearest, best book for beginner, amateur. Simple repairs, raising dropped notes, tuning by easy method of flattened fifths. No previous skills needed. 4 illustrations. 201pp. 5⅜ × 8½. 23267-0 Pa. $3.50

EARLY AMERICAN IRON-ON TRANSFER PATTERNS, edited by Rita Weiss. 75 designs, borders, alphabets, from traditional American sources. 48pp. 8¼ × 11. 23162-3 Pa. $1.95

CROCHETING EDGINGS, edited by Rita Weiss. Over 100 of the best designs for these lovely trims for a host of household items. Complete instructions, illustrations. 48pp. 8¼ × 11. 24031-2 Pa. $2.00

FINGER PLAYS FOR NURSERY AND KINDERGARTEN, Emilie Poulsson. 18 finger plays with music (voice and piano); entertaining, instructive. Counting, nature lore, etc. Victorian classic. 53 illustrations. 80pp. 6½ × 9¼. 22588-7 Pa. $1.95

BOSTON THEN AND NOW, Peter Vanderwarker. Here in 59 side-by-side views are photographic documentations of the city's past and present. 119 photographs. Full captions. 122pp. 8¼ × 11. 24312-5 Pa. $6.95

CROCHETING BEDSPREADS, edited by Rita Weiss. 22 patterns, originally published in three instruction books 1939-41. 39 photos, 8 charts. Instructions. 48pp. 8¼ × 11. 23610-2 Pa. $2.00

HAWTHORNE ON PAINTING, Charles W. Hawthorne. Collected from notes taken by students at famous Cape Cod School; hundreds of direct, personal *apercus*, ideas, suggestions. 91pp. 5⅜ × 8½. 20653-X Pa. $2.50

THERMODYNAMICS, Enrico Fermi. A classic of modern science. Clear, organized treatment of systems, first and second laws, entropy, thermodynamic potentials, etc. Calculus required. 160pp. 5⅜ × 8½. 60361-X Pa. $4.00

TEN BOOKS ON ARCHITECTURE, Vitruvius. The most important book ever written on architecture. Early Roman aesthetics, technology, classical orders, site selection, all other aspects. Morgan translation. 331pp. 5⅜ × 8½. 20645-9 Pa. $5.50

THE CORNELL BREAD BOOK, Clive M. McCay and Jeanette B. McCay. Famed high-protein recipe incorporated into breads, rolls, buns, coffee cakes, pizza, pie crusts, more. Nearly 50 illustrations. 48pp. 8¼ × 11. 23995-0 Pa. $2.00

THE CRAFTSMAN'S HANDBOOK, Cennino Cennini. 15th-century handbook, school of Giotto, explains applying gold, silver leaf; gesso; fresco painting, grinding pigments, etc. 142pp. 6⅛ × 9¼. 20054-X Pa. $3.50

FRANK LLOYD WRIGHT'S FALLINGWATER, Donald Hoffmann. Full story of Wright's masterwork at Bear Run, Pa. 100 photographs of site, construction, and details of completed structure. 112pp. 9¼ × 10. 23671-4 Pa. $6.50

OVAL STAINED GLASS PATTERN BOOK, C. Eaton. 60 new designs framed in shape of an oval. Greater complexity, challenge with sinuous cats, birds, mandalas framed in antique shape. 64pp. 8¼ × 11. 24519-5 Pa. $3.50

THE BOOK OF WOOD CARVING, Charles Marshall Sayers. Still finest book for beginning student. Fundamentals, technique; gives 34 designs, over 34 projects for panels, bookends, mirrors, etc. 33 photos. 118pp. 7¾ × 10⅝. 23654-4 Pa. $3.95

CARVING COUNTRY CHARACTERS, Bill Higginbotham. Expert advice for beginning, advanced carvers on materials, techniques for creating 18 projects—mirthful panorama of American characters. 105 illustrations. 80pp. 8⅜ × 11. 24135-1 Pa. $2.50

300 ART NOUVEAU DESIGNS AND MOTIFS IN FULL COLOR, C.B. Grafton. 44 full-page plates display swirling lines and muted colors typical of Art Nouveau. Borders, frames, panels, cartouches, dingbats, etc. 48pp. 9⅜ × 12¼. 24354-0 Pa. $6.00

SELF-WORKING CARD TRICKS, Karl Fulves. Editor of *Pallbearer* offers 72 tricks that work automatically through nature of card deck. No sleight of hand needed. Often spectacular. 42 illustrations. 113pp. 5⅜ × 8½. 23334-0 Pa. $2.25

CUT AND ASSEMBLE A WESTERN FRONTIER TOWN, Edmund V. Gillon, Jr. Ten authentic full-color buildings on heavy cardboard stock in H-O scale. Sheriff's Office and Jail, Saloon, Wells Fargo, Opera House, others. 48pp. 9¼ × 12¼. 23736-2 Pa. $3.95

CUT AND ASSEMBLE AN EARLY NEW ENGLAND VILLAGE, Edmund V. Gillon, Jr. Printed in full color on heavy cardboard stock. 12 authentic buildings in H-O scale: Adams home in Quincy, Mass., Oliver Wight house in Sturbridge, smithy, store, church, others. 48pp. 9¼ × 12¼. 23536-X Pa. $3.95

THE TALE OF TWO BAD MICE, Beatrix Potter. Tom Thumb and Hunca Munca squeeze out of their hole and go exploring. 27 full-color Potter illustrations. 59pp. 4¼ × 5½. (Available in U.S. only) 23065-1 Pa. $1.50

CARVING FIGURE CARICATURES IN THE OZARK STYLE, Harold L. Enlow. Instructions and illustrations for ten delightful projects, plus general carving instructions. 22 drawings and 47 photographs altogether. 39pp. 8⅜ × 11. 23151-8 Pa. $2.50

A TREASURY OF FLOWER DESIGNS FOR ARTISTS, EMBROIDERERS AND CRAFTSMEN, Susan Gaber. 100 garden favorites lushly rendered by artist for artists, craftsmen, needleworkers. Many form frames, borders. 80pp. 8¼ × 11. 24096-7 Pa. $3.50

CUT & ASSEMBLE A TOY THEATER/THE NUTCRACKER BALLET, Tom Tierney. Model of a complete, full-color production of Tchaikovsky's classic. 6 backdrops, dozens of characters, familiar dance sequences. 32pp. 9⅜ × 12¼. 24194-7 Pa. $4.50

ANIMALS: 1,419 COPYRIGHT-FREE ILLUSTRATIONS OF MAMMALS, BIRDS, FISH, INSECTS, ETC., edited by Jim Harter. Clear wood engravings present, in extremely lifelike poses, over 1,000 species of animals. 284pp. 9 × 12. 23766-4 Pa. $8.95

MORE HAND SHADOWS, Henry Bursill. For those at their 'finger ends," 16 more effects—Shakespeare, a hare, a squirrel, Mr. Punch, and twelve more—each explained by a full-page illustration. Considerable period charm. 30pp. 6½ × 9¼. 21384-6 Pa. $1.95

SURREAL STICKERS AND UNREAL STAMPS, William Rowe. 224 haunting, hilarious stamps on gummed, perforated stock, with images of elephants, geisha girls, George Washington, etc. 16pp. one side. 8¼ × 11. 24371-0 Pa. $3.50

GOURMET KITCHEN LABELS, Ed Sibbett, Jr. 112 full-color labels (4 copies each of 28 designs). Fruit, bread, other culinary motifs. Gummed and perforated. 16pp. 8¼ × 11. 24087-8 Pa. $2.95

PATTERNS AND INSTRUCTIONS FOR CARVING AUTHENTIC BIRDS, H.D. Green. Detailed instructions, 27 diagrams, 85 photographs for carving 15 species of birds so life-like, they'll seem ready to fly! 8¼ × 11. 24222-6 Pa. $2.75

FLATLAND, E.A. Abbott. Science-fiction classic explores life of 2-D being in 3-D world. 16 illustrations. 103pp. 5⅜ × 8. 20001-9 Pa. $2.00

DRIED FLOWERS, Sarah Whitlock and Martha Rankin. Concise, clear, practical guide to dehydration, glycerinizing, pressing plant material, and more. Covers use of silica gel. 12 drawings. 32pp. 5⅜ × 8½. 21802-3 Pa. $1.00

EASY-TO-MAKE CANDLES, Gary V. Guy. Learn how easy it is to make all kinds of decorative candles. Step-by-step instructions. 82 illustrations. 48pp. 8¼ × 11. 23881-4 Pa. $2.50

SUPER STICKERS FOR KIDS, Carolyn Bracken. 128 gummed and perforated full-color stickers: GIRL WANTED, KEEP OUT, BORED OF EDUCATION, X-RATED, COMBAT ZONE, many others. 16pp. 8¼ × 11. 24092-4 Pa. $2.50

CUT AND COLOR PAPER MASKS, Michael Grater. Clowns, animals, funny faces...simply color them in, cut them out, and put them together, and you have 9 paper masks to play with and enjoy. 32pp. 8¼ × 11. 23171-2 Pa. $2.25

A CHRISTMAS CAROL: THE ORIGINAL MANUSCRIPT, Charles Dickens. Clear facsimile of Dickens manuscript, on facing pages with final printed text. 8 illustrations by John Leech, 4 in color on covers. 144pp. 8⅜ × 11¼. 20980-6 Pa. $5.95

CARVING SHOREBIRDS, Harry V. Shourds & Anthony Hillman. 16 full-size patterns (all double-page spreads) for 19 North American shorebirds with step-by-step instructions. 72pp. 9¼ × 12¼. 24287-0 Pa. $4.95

THE GENTLE ART OF MATHEMATICS, Dan Pedoe. Mathematical games, probability, the question of infinity, topology, how the laws of algebra work, problems of irrational numbers, and more. 42 figures. 143pp. 5⅜ × 8½. (EBE) 22949-1 Pa. $3.00

READY-TO-USE DOLLHOUSE WALLPAPER, Katzenbach & Warren, Inc. Stripe, 2 floral stripes, 2 allover florals, polka dot; all in full color. 4 sheets (350 sq. in.) of each, enough for average room. 48pp. 8¼ × 11. 23495-9 Pa. $2.95

MINIATURE IRON-ON TRANSFER PATTERNS FOR DOLLHOUSES, DOLLS, AND SMALL PROJECTS, Rita Weiss and Frank Fontana. Over 100 miniature patterns: rugs, bedspreads, quilts, chair seats, etc. In standard dollhouse size. 48pp. 8¼ × 11. 23741-9 Pa. $1.95

THE DINOSAUR COLORING BOOK, Anthony Rao. 45 renderings of dinosaurs, fossil birds, turtles, other creatures of Mesozoic Era. Scientifically accurate. Captions. 48pp. 8¼ × 11. 24022-3 Pa. $2.25

JAPANESE DESIGN MOTIFS, Matsuya Co. Mon, or heraldic designs. Over 4000 typical, beautiful designs: birds, animals, flowers, swords, fans, geometrics; all beautifully stylized. 213pp. 11⅜ × 8¼. 22874-6 Pa. $6.95

THE TALE OF BENJAMIN BUNNY, Beatrix Potter. Peter Rabbit's cousin coaxes him back into Mr. McGregor's garden for a whole new set of adventures. All 27 full-color illustrations. 59pp. 4¼ × 5½. (Available in U.S. only) 21102-9 Pa. $1.50

THE TALE OF PETER RABBIT AND OTHER FAVORITE STORIES BOXED SET, Beatrix Potter. Seven of Beatrix Potter's best-loved tales including Peter Rabbit in a specially designed, durable boxed set. 4¼ × 5½. Total of 447pp. 158 color illustrations. (Available in U.S. only) 23903-9 Pa. $10.50

PRACTICAL MENTAL MAGIC, Theodore Annemann. Nearly 200 astonishing feats of mental magic revealed in step-by-step detail. Complete advice on staging, patter, etc. Illustrated. 320pp. 5⅜ × 8½. 24426-1 Pa. $5.95

CELEBRATED CASES OF JUDGE DEE (DEE GOONG AN), translated by Robert Van Gulik. Authentic 18th-century Chinese detective novel; Dee and associates solve three interlocked cases. Led to van Gulik's own stories with same characters. Extensive introduction. 9 illustrations. 237pp. 5⅜ × 8½. 23337-5 Pa. $4.50

CUT & FOLD EXTRATERRESTRIAL INVADERS THAT FLY, M. Grater. Stage your own lilliputian space battles.By following the step-by-step instructions and explanatory diagrams you can launch 22 full-color fliers into space. 36pp. 8¼ × 11. 24478-4 Pa. $2.95

CUT & ASSEMBLE VICTORIAN HOUSES, Edmund V. Gillon, Jr. Printed in full color on heavy cardboard stock, 4 authentic Victorian houses in H-O scale: Italian-style Villa, Octagon, Second Empire, Stick Style. 48pp. 9¼ × 12¼. 23849-0 Pa. $3.95

BEST SCIENCE FICTION STORIES OF H.G. WELLS, H.G. Wells. Full novel *The Invisible Man,* plus 17 short stories: "The Crystal Egg," "Aepyornis Island," "The Strange Orchid," etc. 303pp. 5⅜ × 8½. (Available in U.S. only) 21531-8 Pa. $3.95

TRADEMARK DESIGNS OF THE WORLD, Yusaku Kamekura. A lavish collection of nearly 700 trademarks, the work of Wright, Loewy, Klee, Binder, hundreds of others. 160pp. 8¾ × 8. (Available in U.S. only) 24191-2 Pa. $5.00

THE ARTIST'S AND CRAFTSMAN'S GUIDE TO REDUCING, ENLARGING AND TRANSFERRING DESIGNS, Rita Weiss. Discover, reduce, enlarge, transfer designs from any objects to any craft project. 12pp. plus 16 sheets special graph paper. 8¼ × 11. 24142-4 Pa. $3.25

TREASURY OF JAPANESE DESIGNS AND MOTIFS FOR ARTISTS AND CRAFTSMEN, edited by Carol Belanger Grafton. Indispensable collection of 360 traditional Japanese designs and motifs redrawn in clean, crisp black-and-white, copyright-free illustrations. 96pp. 8¼ × 11. 24435-0 Pa. $3.95

KEYBOARD WORKS FOR SOLO INSTRUMENTS, G.F. Handel. 35 neglected works from Handel's vast oeuvre, originally jotted down as improvisations. Includes Eight Great Suites, others. New sequence. 174pp. 9⅜ × 12¼.
24338-9 Pa. $7.50

AMERICAN LEAGUE BASEBALL CARD CLASSICS, Bert Randolph Sugar. 82 stars from 1900s to 60s on facsimile cards. Ruth, Cobb, Mantle, Williams, plus advertising, info, no duplications. Perforated, detachable. 16pp. 8¼ × 11.
24286-2 Pa. $2.95

A TREASURY OF CHARTED DESIGNS FOR NEEDLEWORKERS, Georgia Gorham and Jeanne Warth. 141 charted designs: owl, cat with yarn, tulips, piano, spinning wheel, covered bridge, Victorian house and many others. 48pp. 8¼ × 11.
23558-0 Pa. $1.95

DANISH FLORAL CHARTED DESIGNS, Gerda Bengtsson. Exquisite collection of over 40 different florals: anemone, Iceland poppy, wild fruit, pansies, many others. 45 illustrations. 48pp. 8¼ × 11. 23957-8 Pa. $1.75

OLD PHILADELPHIA IN EARLY PHOTOGRAPHS 1839-1914, Robert F. Looney. 215 photographs: panoramas, street scenes, landmarks, President-elect Lincoln's visit, 1876 Centennial Exposition, much more. 230pp. 8⅜ × 11¾.
23345-6 Pa. $9.95

PRELUDE TO MATHEMATICS, W.W. Sawyer. Noted mathematician's lively, stimulating account of non-Euclidean geometry, matrices, determinants, group theory, other topics. Emphasis on novel, striking aspects. 224pp. 5⅜ × 8½.
24401-6 Pa. $4.50

ADVENTURES WITH A MICROSCOPE, Richard Headstrom. 59 adventures with clothing fibers, protozoa, ferns and lichens, roots and leaves, much more. 142 illustrations. 232pp. 5⅜ × 8½. 23471-1 Pa. $3.50

IDENTIFYING ANIMAL TRACKS: MAMMALS, BIRDS, AND OTHER ANIMALS OF THE EASTERN UNITED STATES, Richard Headstrom. For hunters, naturalists, scouts, nature-lovers. Diagrams of tracks, tips on identification. 128pp. 5⅜ × 8. 24442-3 Pa. $3.50

VICTORIAN FASHIONS AND COSTUMES FROM HARPER'S BAZAR, 1867-1898, edited by Stella Blum. Day costumes, evening wear, sports clothes, shoes, hats, other accessories in over 1,000 detailed engravings. 320pp. 9⅜ × 12¼.
22990-4 Pa. $9.95

EVERYDAY FASHIONS OF THE TWENTIES AS PICTURED IN SEARS AND OTHER CATALOGS, edited by Stella Blum. Actual dress of the Roaring Twenties, with text by Stella Blum. Over 750 illustrations, captions. 156pp. 9 × 12.
24134-3 Pa. $7.95

HALL OF FAME BASEBALL CARDS, edited by Bert Randolph Sugar. Cy Young, Ted Williams, Lou Gehrig, and many other Hall of Fame greats on 92 full-color, detachable reprints of early baseball cards. No duplication of cards with *Classic Baseball Cards*. 16pp. 8¼ × 11. 23624-2 Pa. $2.95

THE ART OF HAND LETTERING, Helm Wotzkow. Course in hand lettering, Roman, Gothic, Italic, Block, Script. Tools, proportions, optical aspects, individual variation. Very quality conscious. Hundreds of specimens. 320pp. 5⅜ × 8½.
21797-3 Pa. $4.95

HOW THE OTHER HALF LIVES, Jacob A. Riis. Journalistic record of filth, degradation, upward drive in New York immigrant slums, shops, around 1900. New edition includes 100 original Riis photos, monuments of early photography. 233pp. 10 × 7⅞. 22012-5 Pa. $7.95

CHINA AND ITS PEOPLE IN EARLY PHOTOGRAPHS, John Thomson. In 200 black-and-white photographs of exceptional quality photographic pioneer Thomson captures the mountains, dwellings, monuments and people of 19th-century China. 272pp. 9⅜ × 12¼. 24393-1 Pa. $12.95

GODEY COSTUME PLATES IN COLOR FOR DECOUPAGE AND FRAMING, edited by Eleanor Hasbrouk Rawlings. 24 full-color engravings depicting 19th-century Parisian haute couture. Printed on one side only. 56pp. 8¼ × 11. 23879-2 Pa. $3.95

ART NOUVEAU STAINED GLASS PATTERN BOOK, Ed Sibbett, Jr. 104 projects using well-known themes of Art Nouveau: swirling forms, florals, peacocks, and sensuous women. 60pp. 8¼ × 11. 23577-7 Pa. $3.00

QUICK AND EASY PATCHWORK ON THE SEWING MACHINE: Susan Aylsworth Murwin and Suzzy Payne. Instructions, diagrams show exactly how to machine sew 12 quilts. 48pp. of templates. 50 figures. 80pp. 8¼ × 11. 23770-2 Pa. $3.50

THE STANDARD BOOK OF QUILT MAKING AND COLLECTING, Marguerite Ickis. Full information, full-sized patterns for making 46 traditional quilts, also 150 other patterns. 483 illustrations. 273pp. 6⅞ × 9⅝. 20582-7 Pa. $5.95

LETTERING AND ALPHABETS, J. Albert Cavanagh. 85 complete alphabets lettered in various styles; instructions for spacing, roughs, brushwork. 121pp. 8¾ × 8. 20053-1 Pa. $3.75

LETTER FORMS: 110 COMPLETE ALPHABETS, Frederick Lambert. 110 sets of capital letters; 16 lower case alphabets; 70 sets of numbers and other symbols. 110pp. 8⅛ × 11. 22872-X Pa. $4.50

ORCHIDS AS HOUSE PLANTS, Rebecca Tyson Northen. Grow cattleyas and many other kinds of orchids—in a window, in a case, or under artificial light. 63 illustrations. 148pp. 5⅜ × 8½. 23261-1 Pa. $2.95

THE MUSHROOM HANDBOOK, Louis C.C. Krieger. Still the best popular handbook. Full descriptions of 259 species, extremely thorough text, poisons, folklore, etc. 32 color plates; 126 other illustrations. 560pp. 5⅜ × 8½. 21861-9 Pa. $8.50

THE DORÉ BIBLE ILLUSTRATIONS, Gustave Doré. All wonderful, detailed plates: Adam and Eve, Flood, Babylon, life of Jesus, etc. Brief King James text with each plate. 241 plates. 241pp. 9 × 12. 23004-X Pa. $6.95

THE BOOK OF KELLS: Selected Plates in Full Color, edited by Blanche Cirker. 32 full-page plates from greatest manuscript-icon of early Middle Ages. Fantastic, mysterious. Publisher's Note. Captions. 32pp. 9¾ × 12¼. 24345-1 Pa. $4.50

THE PERFECT WAGNERITE, George Bernard Shaw. Brilliant criticism of the Ring Cycle, with provocative interpretation of politics, economic theories behind the Ring. 136pp. 5⅜ × 8½. (Available in U.S. only) 21707-8 Pa. $3.00

THE RIME OF THE ANCIENT MARINER, Gustave Doré, S.T. Coleridge. Doré's finest work, 34 plates capture moods, subtleties of poem. Full text. 77pp. 9¼ × 12. 22305-1 Pa. $4.95

SONGS OF INNOCENCE, William Blake. The first and most popular of Blake's famous "Illuminated Books," in a facsimile edition reproducing all 31 brightly colored plates. Additional printed text of each poem. 64pp. 5¼ × 7. 22764-2 Pa. $3.00

AN INTRODUCTION TO INFORMATION THEORY, J.R. Pierce. Second (1980) edition of most impressive non-technical account available. Encoding, entropy, noisy channel, related areas, etc. 320pp. 5⅜ × 8½. 24061-4 Pa. $4.95

THE DIVINE PROPORTION: A STUDY IN MATHEMATICAL BEAUTY, H.E. Huntley. "Divine proportion" or "golden ratio" in poetry, Pascal's triangle, philosophy, psychology, music, mathematical figures, etc. Excellent bridge between science and art. 58 figures. 185pp. 5⅜ × 8½. 22254-3 Pa. $3.95

THE DOVER NEW YORK WALKING GUIDE: From the Battery to Wall Street, Mary J. Shapiro. Superb inexpensive guide to historic buildings and locales in lower Manhattan: Trinity Church, Bowling Green, more. Complete Text; maps. 36 illustrations. 48pp. 3⅞ × 9¼. 24225-0 Pa. $1.75

NEW YORK THEN AND NOW, Edward B. Watson, Edmund V. Gillon, Jr. 83 important Manhattan sites: on facing pages early photographs (1875-1925) and 1976 photos by Gillon. 172 illustrations. 171pp. 9¼ × 10. 23361-8 Pa. $7.95

HISTORIC COSTUME IN PICTURES, Braun & Schneider. Over 1450 costumed figures from dawn of civilization to end of 19th century. English captions. 125 plates. 256pp. 8⅜ × 11¼. 23150-X Pa. $7.50

VICTORIAN AND EDWARDIAN FASHION: A Photographic Survey, Alison Gernsheim. First fashion history completely illustrated by contemporary photographs. Full text plus 235 photos, 1840-1914, in which many celebrities appear. 240pp. 6½ × 9¼. 24205-6 Pa. $6.00

CHARTED CHRISTMAS DESIGNS FOR COUNTED CROSS-STITCH AND OTHER NEEDLECRAFTS, Lindberg Press. Charted designs for 45 beautiful needlecraft projects with many yuletide and wintertime motifs. 48pp. 8¼ × 11. 24356-7 Pa. $1.95

101 FOLK DESIGNS FOR COUNTED CROSS-STITCH AND OTHER NEEDLECRAFTS, Carter Houck. 101 authentic charted folk designs in a wide array of lovely representations with many suggestions for effective use. 48pp. 8¼ × 11. 24369-9 Pa. $1.95

FIVE ACRES AND INDEPENDENCE, Maurice G. Kains. Great back-to-the-land classic explains basics of self-sufficient farming. The one book to get. 95 illustrations. 397pp. 5⅜ × 8½. 20974-1 Pa. $4.95

A MODERN HERBAL, Margaret Grieve. Much the fullest, most exact, most useful compilation of herbal material. Gigantic alphabetical encyclopedia, from aconite to zedoary, gives botanical information, medical properties, folklore, economic uses, and much else. Indispensable to serious reader. 161 illustrations. 888pp. 6½ × 9¼. (Available in U.S. only) 22798-7, 22799-5 Pa., Two-vol. set $16.45

REASON IN ART, George Santayana. Renowned philosopher's provocative, seminal treatment of basis of art in instinct and experience. Volume Four of *The Life of Reason*. 230pp. 5⅜ × 8. 24358-3 Pa. $4.50

LANGUAGE, TRUTH AND LOGIC, Alfred J. Ayer. Famous, clear introduction to Vienna, Cambridge schools of Logical Positivism. Role of philosophy, elimination of metaphysics, nature of analysis, etc. 160pp. 5⅜ × 8½. (USCO) 20010-8 Pa. $2.75

BASIC ELECTRONICS, U.S. Bureau of Naval Personnel. Electron tubes, circuits, antennas, AM, FM, and CW transmission and receiving, etc. 560 illustrations. 567pp. 6½ × 9¼. 21076-6 Pa. $8.95

THE ART DECO STYLE, edited by Theodore Menten. Furniture, jewelry, metalwork, ceramics, fabrics, lighting fixtures, interior decors, exteriors, graphics from pure French sources. Over 400 photographs. 183pp. 8⅜ × 11¼. 22824-X Pa. $6.95

THE FOUR BOOKS OF ARCHITECTURE, Andrea Palladio. 16th-century classic covers classical architectural remains, Renaissance revivals, classical orders, etc. 1738 Ware English edition. 216 plates. 110pp. of text. 9½ × 12¾. 21308-0 Pa. $10.00

THE WIT AND HUMOR OF OSCAR WILDE, edited by Alvin Redman. More than 1000 ripostes, paradoxes, wisecracks: Work is the curse of the drinking classes, I can resist everything except temptations, etc. 258pp. 5⅜ × 8½. (USCO) 20602-5 Pa. $3.50

THE DEVIL'S DICTIONARY, Ambrose Bierce. Barbed, bitter, brilliant witticisms in the form of a dictionary. Best, most ferocious satire America has produced. 145pp. 5⅜ × 8½. 20487-1 Pa. $2.50

ERTÉ'S FASHION DESIGNS, Erté. 210 black-and-white inventions from *Harper's Bazar*, 1918-32, plus 8pp. full-color covers. Captions. 88pp. 9 × 12. 24203-X Pa. $6.50

ERTÉ GRAPHICS, Erté. Collection of striking color graphics: *Seasons, Alphabet, Numerals, Aces* and *Precious Stones*. 50 plates, including 4 on covers. 48pp. 9⅜ × 12¼. 23580-7 Pa. $6.95

PAPER FOLDING FOR BEGINNERS, William D. Murray and Francis J. Rigney. Clearest book for making origami sail boats, roosters, frogs that move legs, etc. 40 projects. More than 275 illustrations. 94pp. 5⅜ × 8½. 20713-7 Pa. $1.95

ORIGAMI FOR THE ENTHUSIAST, John Montroll. Fish, ostrich, peacock, squirrel, rhinoceros, Pegasus, 19 other intricate subjects. Instructions. Diagrams. 128pp. 9 × 12. 23799-0 Pa. $4.95

CROCHETING NOVELTY POT HOLDERS, edited by Linda Macho. 64 useful, whimsical pot holders feature kitchen themes, animals, flowers, other novelties. Surprisingly easy to crochet. Complete instructions. 48pp. 8¼ × 11. 24296-X Pa. $1.95

CROCHETING DOILIES, edited by Rita Weiss. Irish Crochet, Jewel, Star Wheel, Vanity Fair and more. Also luncheon and console sets, runners and centerpieces. 51 illustrations. 48pp. 8¼ × 11. 23424-X Pa. $2.00

YUCATAN BEFORE AND AFTER THE CONQUEST, Diego de Landa. Only significant account of Yucatan written in the early post-Conquest era. Translated by William Gates. Over 120 illustrations. 162pp. 5⅜ × 8½. 23622-6 Pa. $3.50

ORNATE PICTORIAL CALLIGRAPHY, E.A. Lupfer. Complete instructions, over 150 examples help you create magnificent "flourishes" from which beautiful animals and objects gracefully emerge. 8⅛ × 11. 21957-7 Pa. $2.95

DOLLY DINGLE PAPER DOLLS, Grace Drayton. Cute chubby children by same artist who did Campbell Kids. Rare plates from 1910s. 30 paper dolls and over 100 outfits reproduced in full color. 32pp. 9¼ × 12¼. 23711-7 Pa. $2.95

CURIOUS GEORGE PAPER DOLLS IN FULL COLOR, H. A. Rey, Kathy Allert. Naughty little monkey-hero of children's books in two doll figures, plus 48 full-color costumes: pirate, Indian chief, fireman, more. 32pp. 9¼ × 12¼. 24386-9 Pa. $3.50

GERMAN: HOW TO SPEAK AND WRITE IT, Joseph Rosenberg. Like *French, How to Speak and Write It.* Very rich modern course, with a wealth of pictorial material. 330 illustrations. 384pp. 5⅜ × 8½. (USUKO) 20271-2 Pa. $4.75

CATS AND KITTENS: 24 Ready-to-Mail Color Photo Postcards, D. Holby. Handsome collection; feline in a variety of adorable poses. Identifications. 12pp. on postcard stock. 8¼ × 11. 24469-5 Pa. $2.95

MARILYN MONROE PAPER DOLLS, Tom Tierney. 31 full-color designs on heavy stock, from *The Asphalt Jungle, Gentlemen Prefer Blondes,* 22 others. 1 doll. 16 plates. 32pp. 9⅜ × 12¼. 23769-9 Pa. $3.50

FUNDAMENTALS OF LAYOUT, F.H. Wills. All phases of layout design discussed and illustrated in 121 illustrations. Indispensable as student's text or handbook for professional. 124pp. 8⅛ × 11. 21279-3 Pa. $4.50

FANTASTIC SUPER STICKERS, Ed Sibbett, Jr. 75 colorful pressure-sensitive stickers. Peel off and place for a touch of pizzazz: clowns, penguins, teddy bears, etc. Full color. 16pp. 8¼ × 11. 24471-7 Pa. $2.95

LABELS FOR ALL OCCASIONS, Ed Sibbett, Jr. 6 labels each of 16 different designs—baroque, art nouveau, art deco, Pennsylvania Dutch, etc.—in full color. 24pp. 8¼ × 11. 23688-9 Pa. $2.95

HOW TO CALCULATE QUICKLY: RAPID METHODS IN BASIC MATHEMATICS, Henry Sticker. Addition, subtraction, multiplication, division, checks, etc. More than 8000 problems, solutions. 185pp. 5 × 7¼. 20295-X Pa. $2.95

THE CAT COLORING BOOK, Karen Baldauski. Handsome, realistic renderings of 40 splendid felines, from American shorthair to exotic types. 44 plates. Captions. 48pp. 8¼ × 11. 24011-8 Pa. $2.25

THE TALE OF PETER RABBIT, Beatrix Potter. The inimitable Peter's terrifying adventure in Mr. McGregor's garden, with all 27 wonderful, full-color Potter illustrations. 55pp. 4¼ × 5½. (Available in U.S. only) 22827-4 Pa. $1.50

BASIC ELECTRICITY, U.S. Bureau of Naval Personnel. Batteries, circuits, conductors, AC and DC, inductance and capacitance, generators, motors, transformers, amplifiers, etc. 349 illustrations. 448pp. 6½ × 9¼. 20973-3 Pa. $7.95